THE SKYLIGHT BOOK

Capturing the Sun and the Moon,
A Guide to Creating Natural Light

Running Press

THE SKYLIGHT BOOK

Capturing the Sun and the Moon,
A Guide to Creating Natural Light

by Al Burns

Running Press
Philadelphia, Pennsylvania

Printed in the United States of America

Distributed in Canada by Van Nostrand Reinhold Ltd., Ontario

2 3 4 5 6 7 8 9

First digit on left indicates the number of this printing.

Library of Congress Cataloging in Publication Data
Burns, Al, 1945-
The Skylight Book.

Bibliography: p.
1. Skylights—Handbooks, manuals, etc.
I. Title.
TH2487.B87 690'.1'5 76-40974
ISBN 0-914294-55-5 Library binding
ISBN 0-914294-56-3 Paperback

Running Press
38 South Nineteenth Street
Philadelphia, Pennsylvania 19103

Asphalt and cedar shing

Red line must line-up with surface of sheathing in order to make sure head cladding and flashing are correctly i locked. See detail 2.
For thicker roofing material (shakes) red line should be approx. 5/8" above sheathing. See also detail 'A' under s

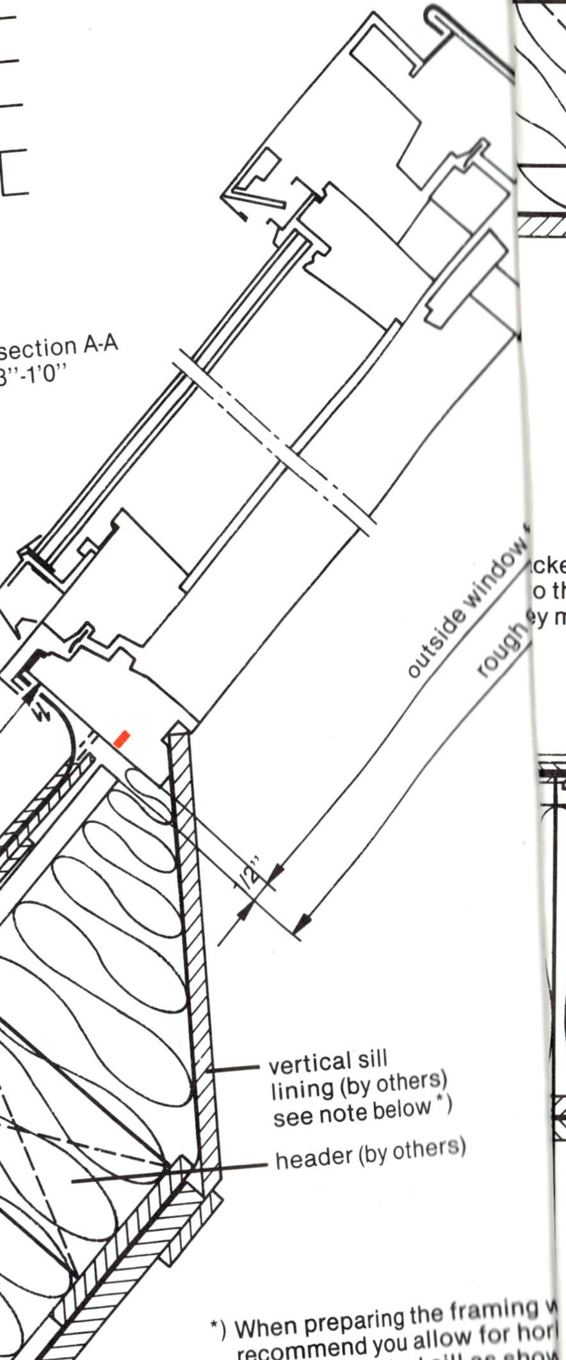

section A-A
3"-1'0"

outside window
rough

1/2"

vertical sill lining (by others) see note below *)

header (by others)

*) When preparing the framing w recommend you allow for hori soffit and vertical sill as show this drawing. It allows for bet distribution of daylight.

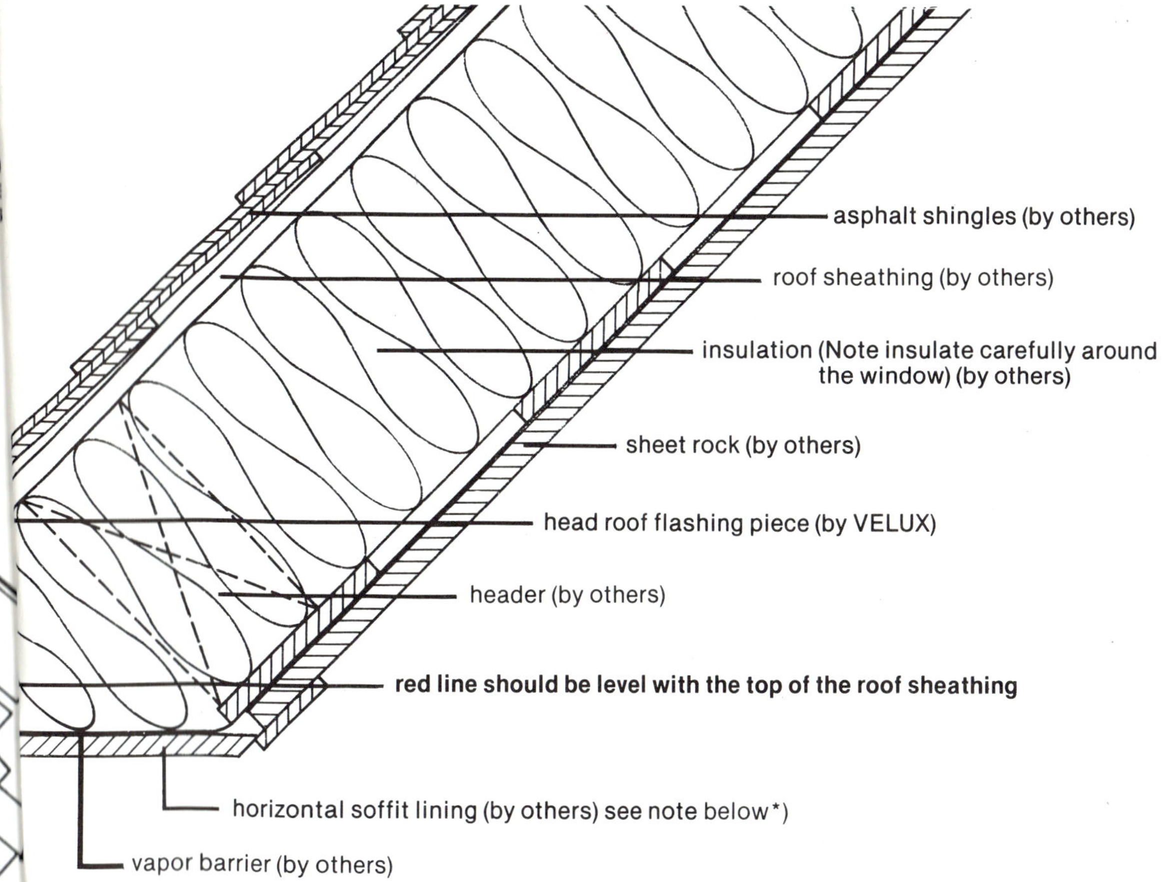

asphalt shingles (by others)

roof sheathing (by others)

insulation (Note insulate carefully around the window) (by others)

sheet rock (by others)

head roof flashing piece (by VELUX)

header (by others)

red line should be level with the top of the roof sheathing

horizontal soffit lining (by others) see note below*)

vapor barrier (by others)

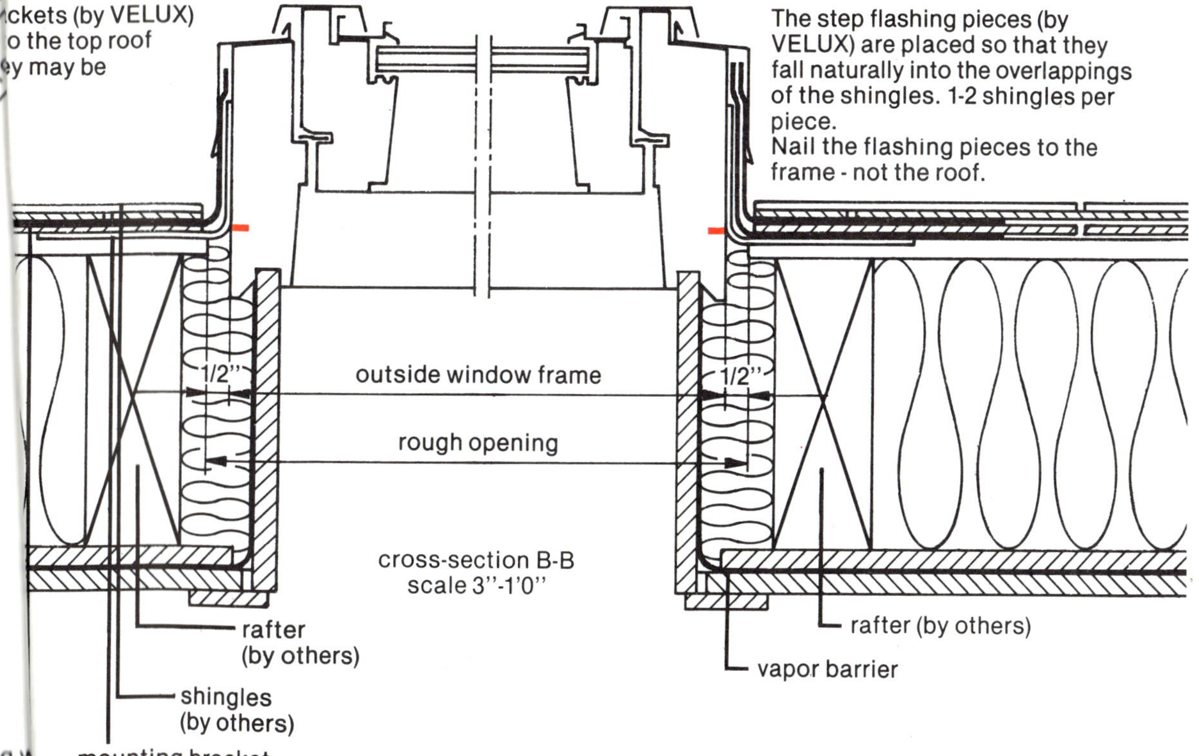

ckets (by VELUX)
o the top roof
y may be

The step flashing pieces (by VELUX) are placed so that they fall naturally into the overlappings of the shingles. 1-2 shingles per piece.
Nail the flashing pieces to the frame - not the roof.

1/2"

outside window frame

1/2"

rough opening

cross-section B-B
scale 3"-1'0"

rafter (by others)

shingles (by others)

mounting bracket (by VELUX)

shing
UX)

rafter (by others)

vapor barrier

VELUX-AMERICA INC.
74 Cummings Park, Woburn,
Massachusetts 01801. Tel. (617) 935-7390, 935-7848

Note!
You save time and money by reading before installing.

Installation of window type GGL using VELUX flashing type L.

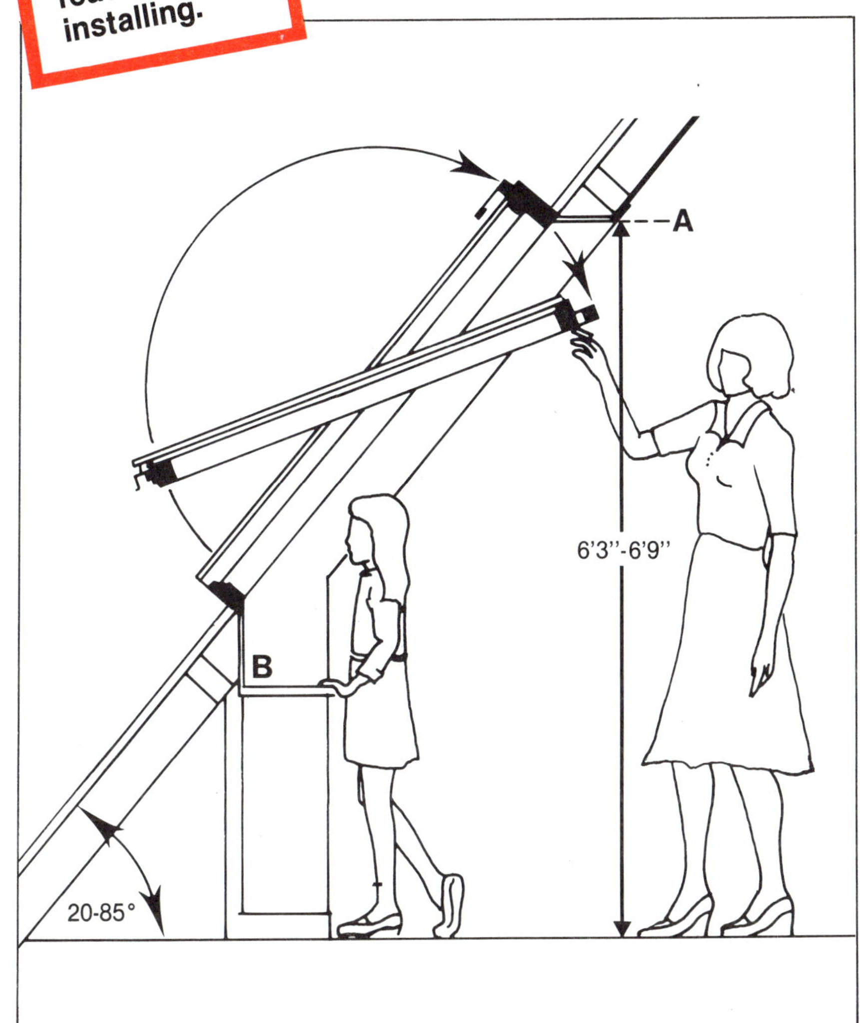

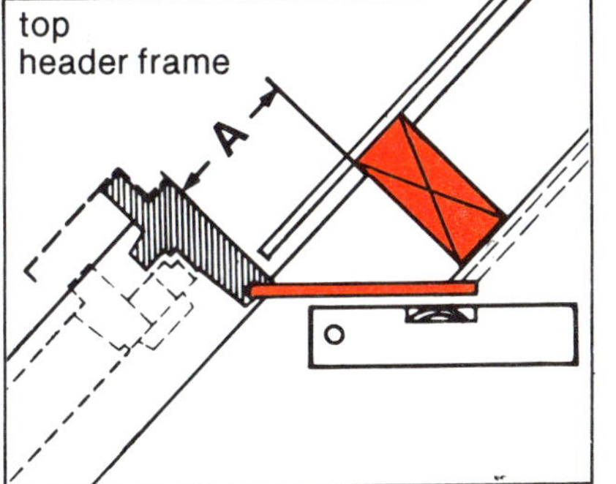

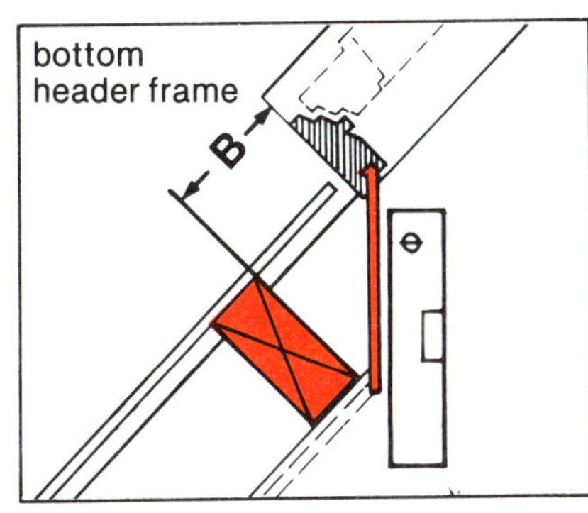

Operation height

For comfortable operation by hand the height from finished floor level to the top of window should be 6'3"-6'9". Though not essential, a horizontal soffit at the head (A) and vertical lining at the sill (B) will give more headroom, better light distribution and an attractive appearance. Manual and remote power controls available for out-of-reach operation.

Heading of the rafters

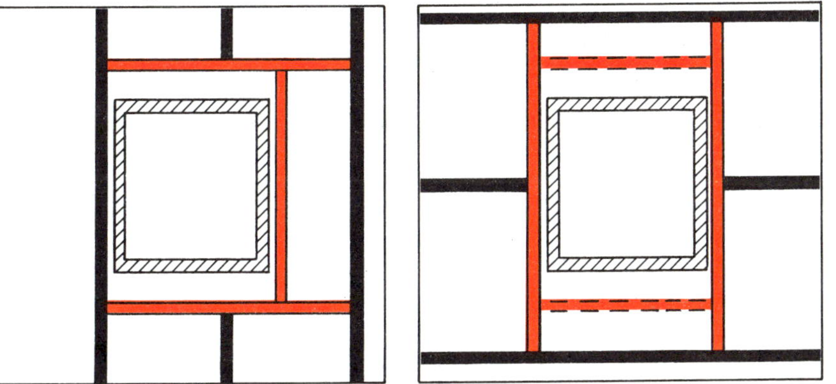

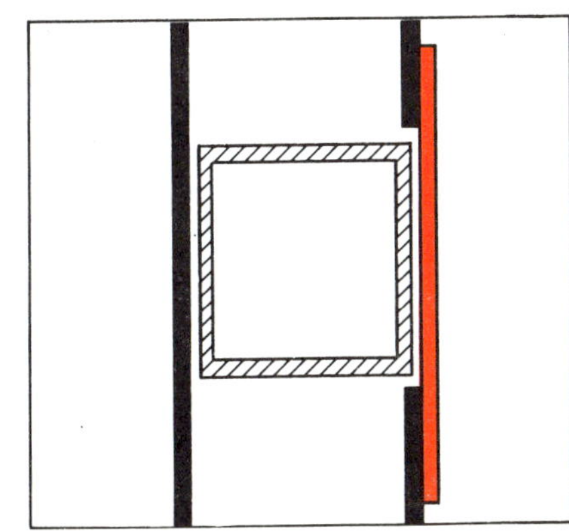

Prepare and install window frame.

1.
Place window on side frame
Separate sash from frame by turning the release screw 'S' all the way clockwise.

2.
Remove the four exterior alum. cladding pieces from the frame. Note: 2 and 4 are not interchangeable.
Do not remove 5 and 6.

3.
Screw the mounting brackets to the frame at the red line as shown below.

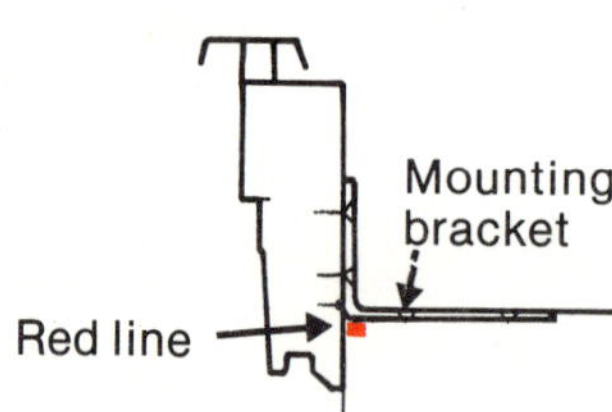

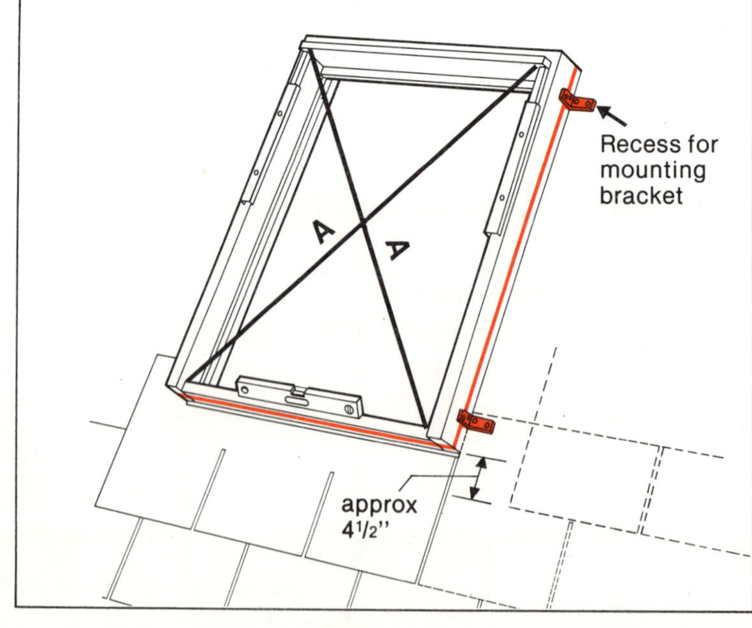

4.
Install the frame in the roof. Screw the Mounting brackets into the sheathing. Red line on frame must line-up with surface of sheathing. Level the sill and take diagonal measures to assure squareness.

Install VELUX prefabricated roof flashing.

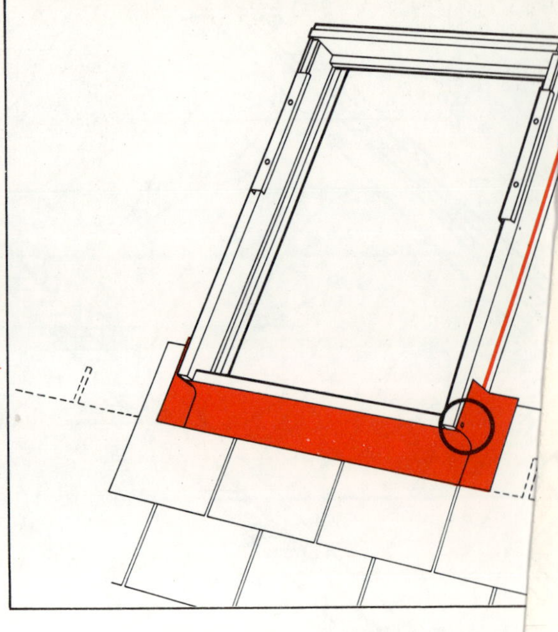

5.
Install the sill flashing section. It should be even with the front edge of the shingles. Nail the side pieces to the frame.

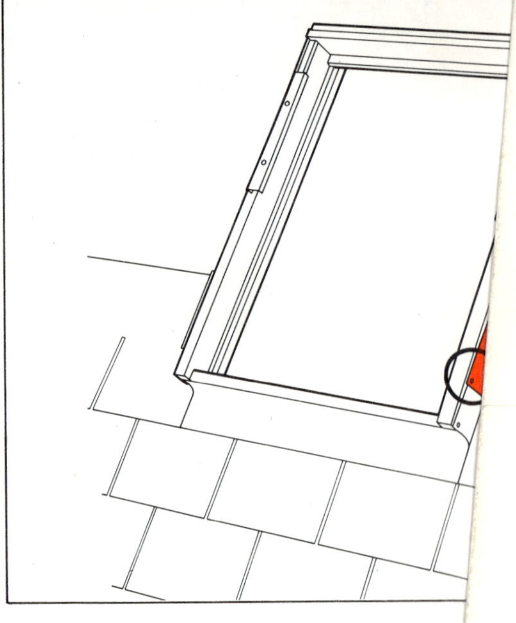

6.
Install the step flashing pieces. They should fall naturally into the overlapping of the shingles. 1-2 shingles per piece.
Note: Nail the pieces to the frame as shown in detail.

7.
Install the top step flashing piece. When necessary cut the piece as shown in detail.

Replace exterior claddings and sash.

8.
Replace exterior claddings. Remember pieces 2 and 4 are not interchangeable.

9.
Install the head flashing section. Interlock with side claddings 2 and 4. Do not nail to roof. See detail A + B for use of filler piece.

A For thick cedar shakes use filler piece supplied.

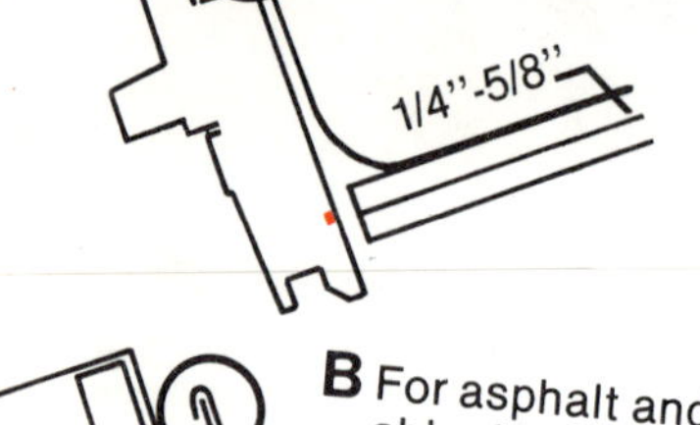

B For asphalt and cedar shingles filler piece unnecessary.

0-1/4"

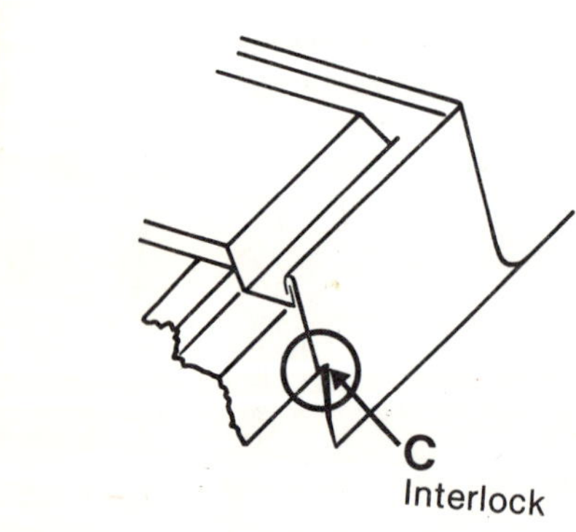

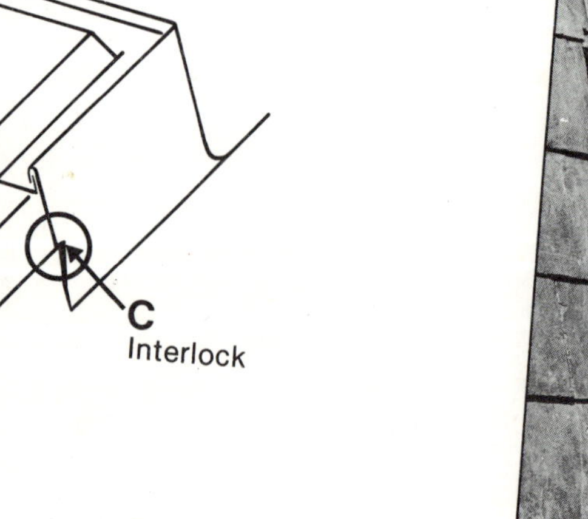

10.
Replace sash in frame. Make sure release screw 'S' is turned all the way out. (Counter-clockwise).

VELUX-AMERICA INC. 74 Cummings Park, Wobur

Author's Premise

This book is about letting the universe into your home as economically and sensibly as possible.

It would be well perhaps if we were to spend more of our days and nights without any obstruction between us and the celestial bodies.

—Henry David Thoreau

Art direction and interior design by Fetterman and Fetterman
Cover design by Jim Wilson
Cover illustration by Gil Cohen
Cover photography and interior photography by Urs Hilfiker
Additional interior photography by Jill Carmody-Burns

Special thanks to Richard E. Nicholls, The Plant Doctor, for the section on "Plants Under Your Skylight"

Typography: Stymie light, by Type Design Innovations, Philadelphia
Cover printed by Harrison Color Process Lithographers, Willow Grove, Pa.
Printed and bound by Port City Press, Baltimore

Topics

Some Little Thoughts About Big Things

So you want to bring a piece of the sky into your home. And not just the big blue, but the deep black, speckled with silver, and a chunk of the moon as well. Watch the Big Dipper move over your bed, then twist out of view, and enjoy feeling close with the natural cycles of dark and light. Really, it seems that there *ought* to be more skylights. And, in fact, it's all quite easy to bring about.

In function, if not form, the modern skylight could be said to date back to the ancient *clerestory.* The clerestory ("clere" in the sense of lit up) is a high-rising structure, pierced by windows, designed for admitting greater light (and sometimes air) into the central body of the main structure. Some of the great Egyptian temples were known for these roofs-above-the-roof. Modified by the Greeks and Romans over the ages, the clerestory became a characteristic element of medieval Gothic architecture. The Cathedral of Nôtre Dame in Paris shows the typical structure: the nave of the church receives the grace of light through the stained-glass windows in the clerestories rising above the roofs of the cathedral's outer aisles.

Unfortunately, perhaps, most of us don't live in Gothic cathedrals. For many people around the globe, a well insulated, protective shelter has always been a basic need. And not until fairly recently in history has the technology of materials made it practical for individuals to feel the sky close within their homes.

Just when a piece of glass was first used over a hole in a roof—and who put it there—I couldn't say. Probably it was not too long after glass side windows became common. By about the year 1 AD transparent glass sheets were first made in Rome. Glass manufacture came to be dominated by the Venetians throughout the Middle Ages. In France, the manufacture of plate glass in 1688 led to the widespread use of mirrors and larger and stronger glass windows. Although plate glass did not win popularity in America until 1850, window glass—made by spinning flat a bubble of blown glass—had been in great demand since the early part of the nineteenth century.

Acrylic plastic, commonly known as Plexiglas, is perhaps a modern-day counterpart to the translucent, transparent material developed by our ancestors. The contemporary surge of interest in skylight construction owes much to the plastics industry. Now the materials are readily accessible. And it's left for each of us, through a piece of well-placed glass, to bring the sun and moon-lit sky within our everyday vision.

Maybe we no longer need Thoreau to remind us that "It would be well perhaps if we were to spend more of our days and nights without any obstruction between us and the celestial bodies" *(Walden).* Yet we might find reinforcement in Thoreau's belief that the barrier stands to be broken by the simplest and most enjoyable means: "Shall we forever resign the pleasure of construction to the carpenter? What does architecture amount to in the experience of the mass of men? I never in all my walks came across a man engaged in so simple and natural an occupation as building his house" (*Walden* again).

Some part of that pleasure will be yours when you cut a hole in your roof and set down the glass. And here *is* an expert carpenter to tell anyone the best way to do it.

—THE EDITOR

Planning for Your Skylight

To get the most out of this book once you have decided to build your skylight, you should read it through *entirely* before you begin the work. Some of the different kinds of work involved do overlap, so it's a good idea to know before you start *all* that needs to be done during any one phase of construction. If you are not quite sure whether you really want to build your own skylight, then reading this section may help you make up your mind.

There are a lot of different types of skylights and a lot of different places to put them in, so we should consider as many situations as possible for both present and future building. Planning a skylight which answers the requirements of both your desires and your budget is not as difficult as it may seem. Innovations in new building materials have enabled the weekend carpenter and designer to really create and change his home environment.

There are many different materials that can be used as a roof covering. Wood, asphalt and asbestos shingles, tile or slate, and materials like tin, aluminum, copper, and galvanized iron are the most common materials used for a pitched roof. The covering generally used for a flat or low-pitched roof is built up layers of roofing-felt with a gravel topping.

Of all these roof coverings, the slate and tile roofs are the only ones I would not recommend for you to build a skylight in, without some more investigation. You will be able to build a skylight using the curb and flashing method with no difficulty in all other roof coverings.

Start with evaluating what you want from a skylight. Remember, they are called skylights, not sunlights. They let in more than just the sun. A full moon rising in your

AN OLD WAREHOUSE CONVERTED TO LIVING SPACE BY DECKER & BOROWSKY, PHILADELPHIA ARCHITECTURAL FIRM. THE SKYLIGHT WAS INSTALLED TO FULFILL ONE OF THE TRADITIONAL FUNCTIONS, LIGHTING THE STAIRWELL.

A SMALL BATHROOM ENLARGED BY THE ADDITION OF A SMALL SKYLIGHT.

skylight is really a very special sight. Even if your home is covered in shade, you will be surprised at the amount of light that a skylight will let in. A skylight over a dark stairway or a gloomy corner is a natural. If you feel your attic bedroom is too small, put in a skylight—and let in the universe.

One place where a skylight is especially effective is the bathroom. Fill the skylight with plants and take a lot of baths and showers. The plants will freak out with all that warm air and humidity, along with the sunlight. The one thing not to worry about is whether the skylight is in a bad place—no matter where you decide to put it, you are bound to be pleased with the result. If you have a situation where the skylight has to go through a ceiling as well as the roof, again, don't worry—there is a real easy way to frame in that ceiling and finish it in a variety of attractive ways.

If your roof has a leak in it, you should *not* fix the leak by cutting it out and installing a skylight. The reason you should not do this is that sometimes the place where the leak shows itself on the inside may not be the place where it is on the outside. Because water follows the course of least resistance, it could run down the rafters for a great distance before it shows itself internally.

The proper thing to do is first determine where the water enters the house. Look for tears in the covering or for nails that have popped up. Check very carefully around the flashings of the chimney and vents which protrude from the roof. If you determine after a careful study that the leak does originate from an area where you wish to put a skylight, then go ahead and build one.

After deciding where you are going to put a skylight, figure out how much you want to spend. To help in the figuring, break down the whole job of erecting the skylight into small projects, each of which has its own bulk of materials. First there is the erection of the curb. If you are installing a ready-made skylight, consult the instructions that are provided with it. The curb and all the materials related to its construction should run about \$25 to \$35 for a 4′ x 4′ skylight. This curb is needed for all the skylights shown in this book and is used for most commercially manufactured skylights as well. The second step is to figure which type of skylight to use. The manufactured variety is expensive, somewhere between three and four times as much as your own would be. For example, you can purchase a 24″ x 24″ prefabricated skylight for about \$125.00; for the same price you can put in a *complete* 48″ x 48″ skylight, curb included, and probably still have enough left to buy beer for the help.

As you can see, your money goes a lot further in terms of size and creativity if you build your own.

There are two styles of skylights that I recommend for the home builder. These are the Plexiglas, or plate glass, skylight, and the Plexiglas box skylight. The reason that I recommend these two is twofold: the materials involved are standard and easy to obtain almost anywhere, and the construction methods are practical and easy for the weekend carpenter to deal with. If in reading this book you come up with what you think is a better way, then do it.

> If there is any danger of something dropping onto the skylight, like a limb from an overhanging tree, then in my opinion Plexiglas is the material you should use.

The Plexiglas, or plate glass, skylight is probably the most common of the skylights. Here the actual skylight material is laid along the top edge of the curb and *aluminum angle* is placed over this edge to make it watertight. To insure a watertight joint, a bead of *silicone caulk* is laid first between the Plexiglas and the curb and then between the Plexiglas and the aluminum angle. For most of us this will be the best type to use. The advantage of this type exists in the large variety of materials which can be used for the skylight. Clear or tinted Plexiglas, or the white Plexiglas which allows light to enter but cannot be seen through, can be used. If you need the safety of a high-strength material you can install *wire glass,* which can be purchased at a good glass store. Use your Yellow Pages.

The box-frame skylight was created as another solution to the problem of making the joint between the skylight and the curb watertight. The box-frame skylight is built to fit over the top of the curb; its sides extend down over the exterior sides of the curb, thereby eliminating any need for a top flashing. In solving this problem very well, however, a whole new process has to be learned (by most of us)—that of cementing Plexiglas. It is not difficult to do and some will even find it easier to make the box-frame than to work with aluminum angle. Both of these are good solutions to the same problem.

The cost of the Plexiglas for the box skylight, or for the Plexiglas or plate glass style, is about \$60 for a 4′ x 4′ piece. The wire-glass material is in the same price range.

A CONTEMPORARY KITCHEN UTILIZING NATURAL LIGHT TO ENHANCE THE WORKING ENVIRONMENT.

STUDIO SPACE WITH A PARTICULARLY LARGE CENTRALLY PLACED SKYLIGHT.

The aluminum angle that is used to seal and hold the skylight material in place can be bought in any good hardware store. It will cost about $10 to $15. For caulking, I recommend very strongly that you spend the extra money and buy silicone caulk made by either DuPont or G.E. The cost of this product is worthwhile, considering what it does and how long it will last.

The last step in designing and estimating the cost is determining how the inside of the skylight will look. "Finishing" the skylight should be thought out very carefully. Because of all of the possibilities, you should take your time to decide what you want. If the ceiling of the room is the roof of the house, like in an attic room or top room where the ceiling has been removed, then there is not much for you to do in the way of finishing the skylight. If, on the other hand, you have to cut through both a roof and a ceiling, then trimming your skylight is very important. The first thing is to frame out the hole and then cover it to go with the rest of the ceiling. What you are doing essentially is creating a *shaft* for the skylight. Lights can be added in the shaft as a pleasant way to bring diffused light into a room.

All sorts of materials can be used to cover the walls of the shaft. Drywall can be installed to match the ceiling. If plants are going in the shaft you might want to put some wood paneling there. Or maybe cover the walls in burlap. Whatever you choose to do will look real good.

The most expensive material you will have to purchase is the framing material. The total cost will probably be about $15 to $25 for trimming. Of course, if you use an expensive wall covering or have lights in the shaft, the price will go up. Because there is not so much area in the shaft, some wall coverings that are prohibitive costwise for large areas may do just right here.

Whichever type of skylight you choose, you will need a plan of some sort. If you are doing the skylight yourself, then the plan need not be any more than a rough sketch. Just something to help you picture it. On the other hand, if you are having a skylight built by someone else, you ought to have a pretty firm idea of what you want, regarding both the skylight itself and how it should be trimmed. In your discussions with the builder, everything should be resolved before the work is begun. This will save you money in the long run by avoiding those unanticipated changes which always manage to increase your cost. And if you are considering buying a manufactured skylight—although I don't recommend them—you still have to know what to look for in the product; check out the little section on manufacturers at the end of this book.

One further note: *thermal glass* is an option for building the Plexiglas or plate glass styles of skylight; thermal glass will keep heat loss at a minimum. *However,* because of the cost factor, and because there is another way to insulate the skylight itself, I don't recommend the use of thermal glass. But if extra insulation is desired, the section on internal storm windows will tell you all you need to know.

PRECEDING PAGES

THE AUTHOR'S BATHROOM SKYLIGHT BUILT WITH A FOUND WINDOW.

TWO 4' X 8' SKYLIGHTS RUN THE FULL LENGTH OF THIS BEDROOM.

AN 18TH-CENTURY CHESTER COUNTY (PENNSYLVANIA) STONE FARMHOUSE WITH TWO HUGE ACRYLIC DOME SKYLIGHTS PLACED PERFECTLY OVER BOTH SLEEPING AREAS. THIRD PHOTO SHOWS ACCESS TO SLEEPING LOFT PICTURED IN FIRST PHOTO.

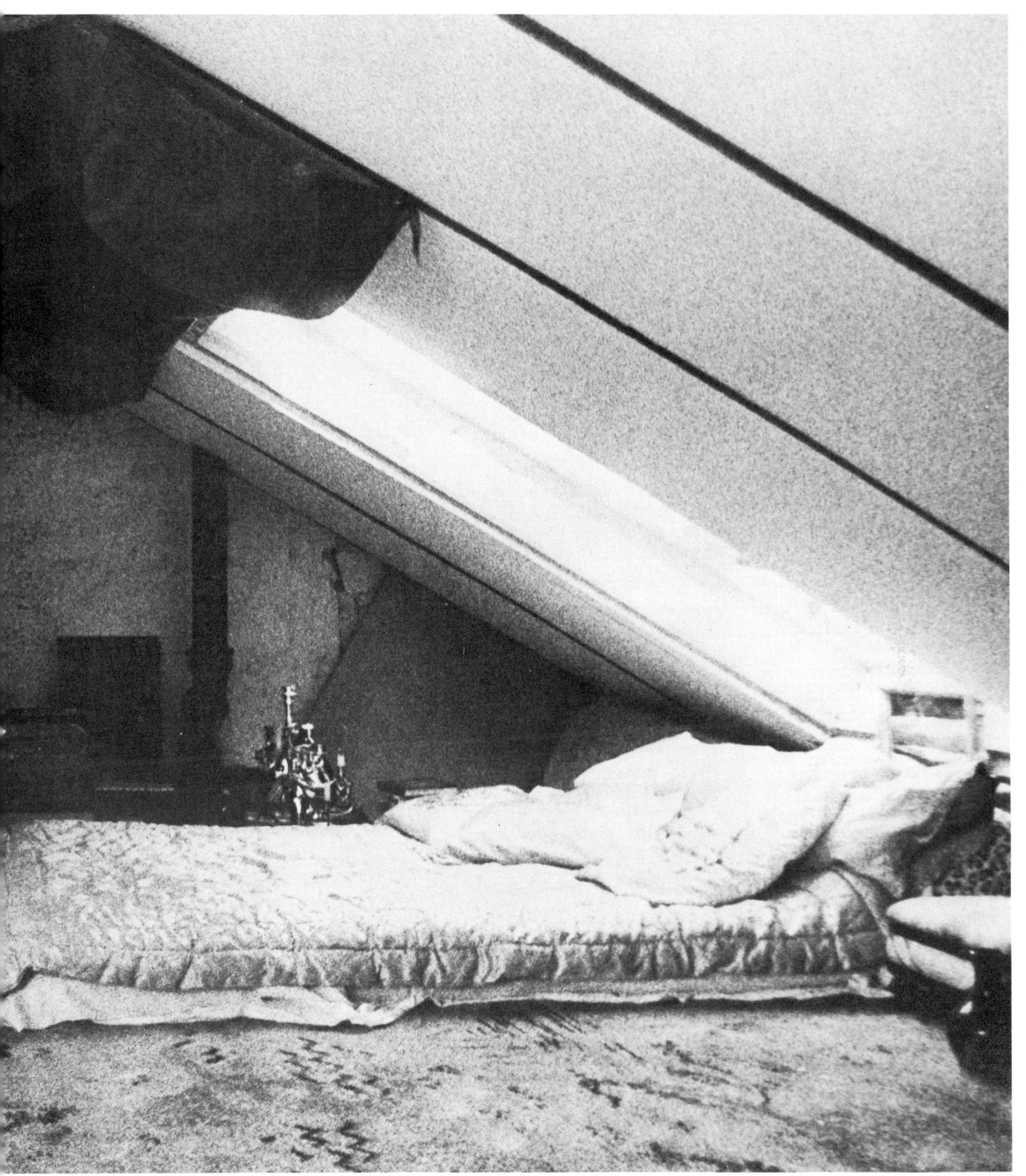

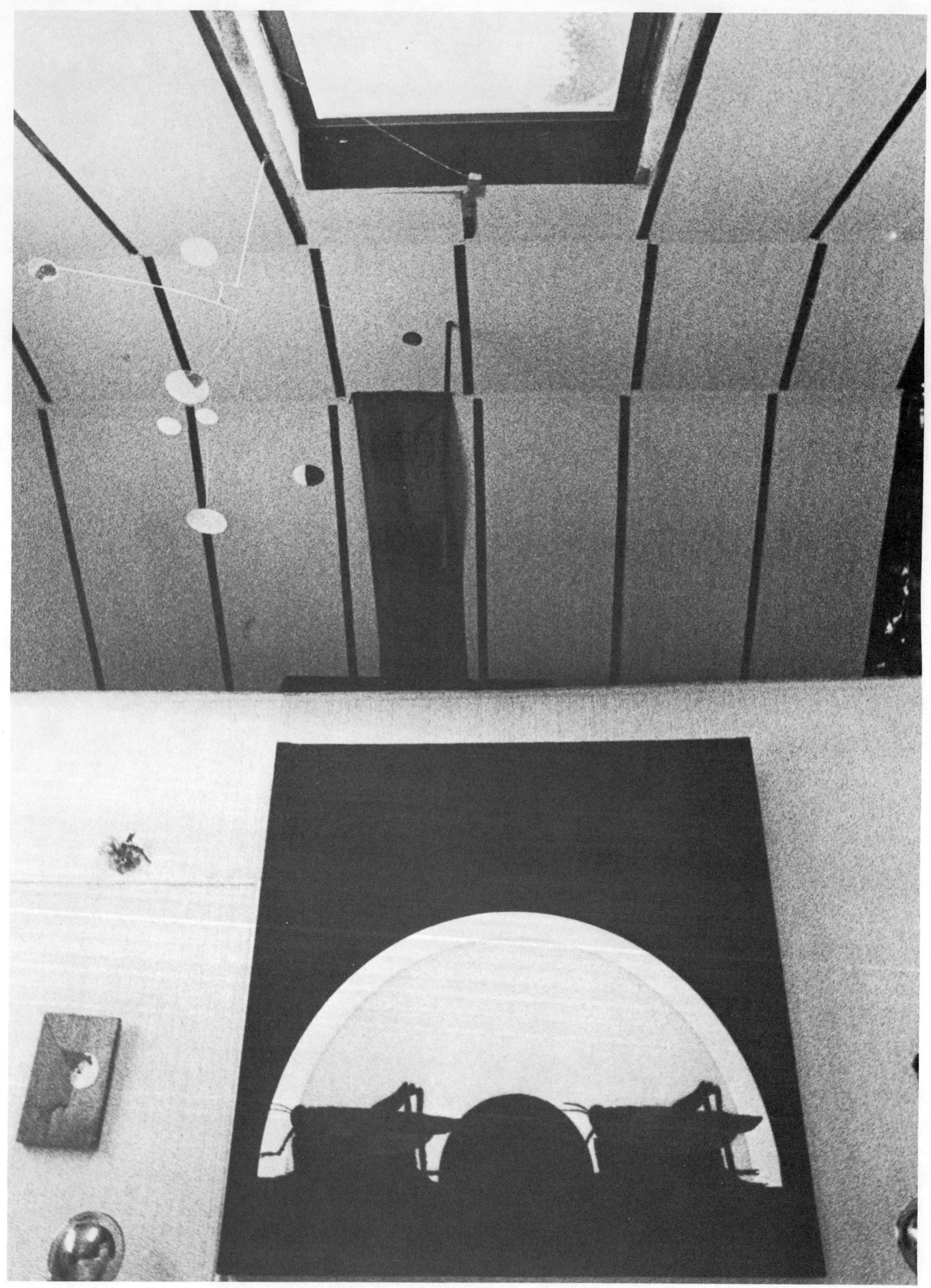

A DECKER & BOROWSKY (THE PHILADELPHIA ARCHITECTURAL FIRM) LOFT RENOVATION EMPLOYING SEVERAL SKYLIGHT LOCATIONS TO ADD NATURAL LIGHT TO WIDE OPEN SPACES.

PRECEDING PAGES

A PHILADELPHIA CARRIAGE HOUSE WITH THE ORIGINAL SKYLIGHT INTACT. THE OWNER-BUILDER ADDED A FRAME STRUCTURE FOR HANGING PLANTS.

A VIEW FROM THE BED LOOKING UP INTO A TWO-STORY LIGHT SHAFT IN THE EYES GALLERY, PHILADELPHIA

FOLLOWING

AN EXISTING SKYLIGHT WITH THE SUPPORTING STRUCTURE EXPOSED TO ALLOW ADDITIONAL LIGHT PENETRATION.

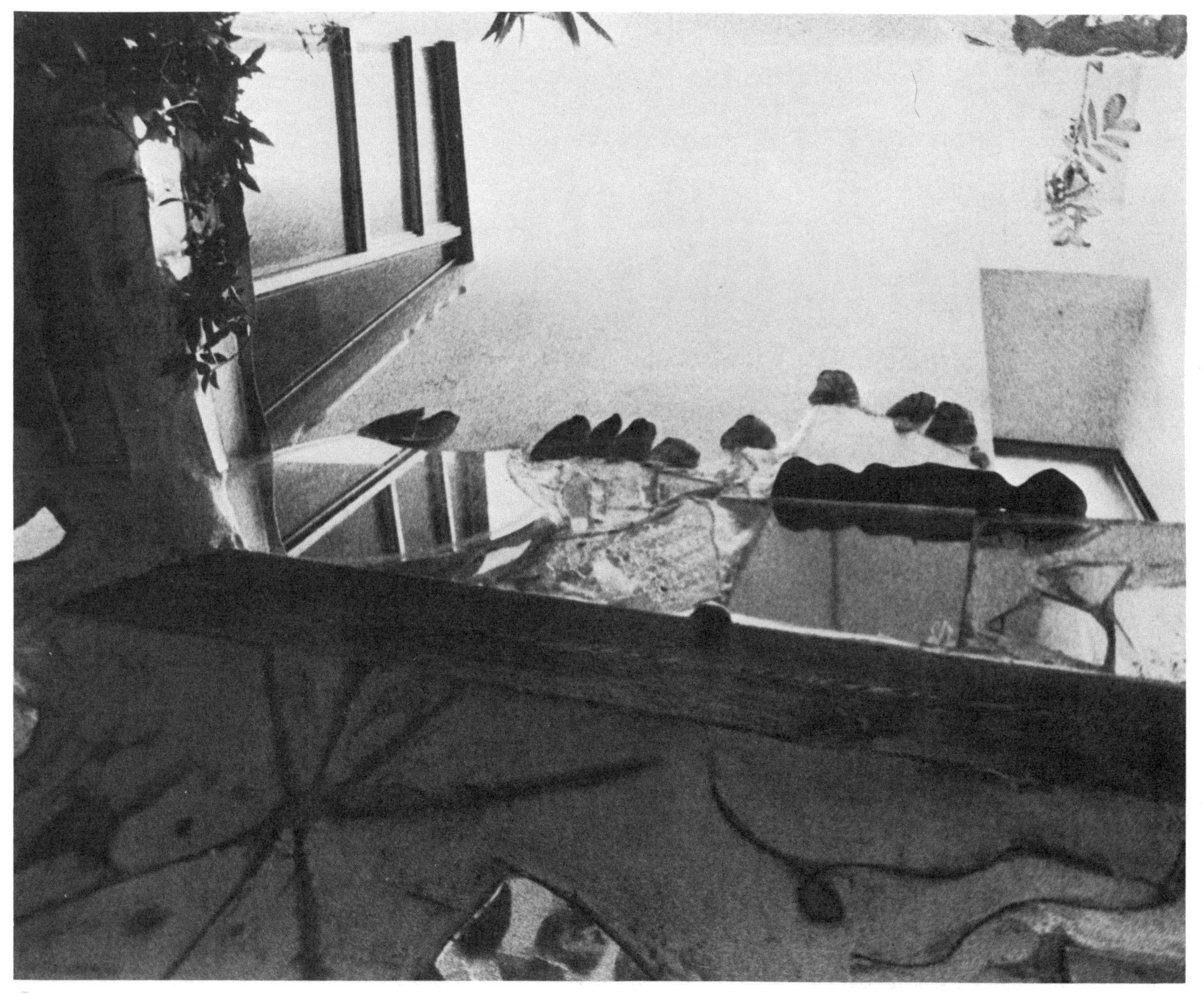

PRECEDING PAGES

A SKYLIGHT ENHANCES THIS ENVIRONMENT, CREATED AS A STUDY IN TEXTURES.

LOOKING UP FROM THE BOTTOM OF A STAIRWAY IN THE WORKS GALLERY (PHILADELPHIA) AT A STAINED GLASS SKYLIGHT. THE STAINED GLASS REPLACED A LESS DRAMATIC INDUSTRIAL SKYLIGHT.

THIS HALLWAY USED TO BE DARK.

SKYLIGHTS ARE PERFECT FOR BRINGING OUT THE DYNAMIC EFFECTS OF STAINED GLASS. (THIS ROOM IS ALSO PICTURED ON THE BACK COVER.)

FOLLOWING PAGES

A LARGE COMMERCIAL ACRYLIC DOME, FACING SOUTH TO RECEIVE HEAT. AN OUTSIDE ADJUSTABLE SHADE COVERS IT IN SUMMER.

YOU ARE LOOKING THROUGH AN OPEN CENTRAL SHAFT OF THIS OWNER-BUILT HOME IN THE WOODS. THE CENTRAL SKYLIGHT FLANKED ON TWO SIDES BY LARGE ROOF WINDOWS PROVIDES INCREDIBLE NATURAL LIGHT, AS WELL AS A FINE AWARENESS OF THE TREES AND SKY.

A GEODESIC SKYLIGHT BUILT OF PLEXIGLAS AND WOOD GIVES THIS CITY STUDIO THE ADDED DIMENSION OF HEIGHT.

TWO SKYLIGHTS IN TWO ROOMS BUILT BY THE AUTHOR FOR THE PUBLISHER. LOCATED DIRECTLY OVER EACH BED, THEY SURE MAKE WAKING UP A LOT NICER.

AMERICA

A SCULPTOR'S PROBLEM OF FINDING NATURAL LIGHT IS SOLVED BY A HOLE IN THE ROOF.

Essential Tools: Hand and Power

Before you go any further you should take some time and get your tools in order. Everyone has most of the basic hand tools for doing small carpentry jobs around the house, but a quick check here will tell you if your collection is complete enough to handle the skylight job. Here is a list of the minimum hand tools needed:

Hammer — A good 16 oz. hammer with a straight or curved claw.

Screwdrivers — two straight blade: one large 8", one smaller 4"; and one Phillips screwdriver with a #2 point.

Chisels — One with a ¾" blade is sufficient here.

Saw — A good hand saw is very important. It could make the difference between an enjoyable job and an unpleasant one. Hand saws are measured by the teeth (called points) per inch. An 8-point saw is a good first hand saw.

Ruler — An 8- or 12-foot metal tape measure is your best bet for a reliable tool.

Utility or mat knife — For the money (under $2 for the best), it is probably your most used tool. From sharpening pencils to cutting roofing and flashings, you will be using it constantly.

Framing square — If you do any kind of building, get a good square. Aluminum is the best because it will never rust and is very durable. You can use this tool for laying out studs, drawing straight lines, or cutting drywall.

Flooring chisel or electrician's chisel — This is a really valuable tool for cutting sheet metal or prying things like molding and door jambs. It is real good for cutting away the roof material. The blade is about 2½" wide.

Chalk line — Used for marking lines. There are two types. With one kind you apply the chalk to the string; the chalk comes in a small block and is run along the string. The other is a self-contained unit which has the string on a reel; powdered chalk is inside the unit so that when the string is pulled out the chalk is applied automatically. This unit may be used as a plumb bob also.

Plumb bob — Weighted, pointed object with a string attached to one end. It is used for marking vertical lines.

Although there are many more, these are the basic hand tools with which almost any carpenter's job can be done. The only piece of advice I would like to give to someone buying hand tools is DON'T BUY CHEAP. A cheap hand tool is nothing but a headache and a safety hazard. The cheap hammer whose head pulls off or, worse yet, flies off when you are nailing is something none of us need to contend with. Or the screwdriver that twists out of shape before we have screwed in five screws is a waste of time. Buy good hand tools and they will last a lifetime. The best and easiest way to buy good hand tools is to go to a good hardware store and ask. They are one of your best sources of information.

Power tools. One difference between power and hand tools is that, with power tools, cheap does *not* necessarily mean bad, as it does with the hand-operated ones.

Using price somewhat as a guideline, you can pretty well figure out what you are buying. But there are some other things to look for in trying to determine the quality of the tool.

Amps versus horsepower. Every electric motor is rated in amperage; the larger the amperage the better the motor. If the tool is listed only in horsepower, then the tool should be suspect: the reason that the amps are not listed is that they are lower than they should be. In other words, although a motor may have what seems like a good horsepower rating, it may result in a premature burn-out if the amp rating does not correspond.

The amp rating varies for different tools and for different uses for those tools. The chart below will help in figuring the correct horsepower and amp rating. Another clue of a tool's quality is the type of bearing in the motor. There are needle and roller bearings and bronze bushings. The needle and roller bearings are much better than the bronze bushings, but the bronze bearings will work okay if they are kept well oiled.

When you get set to buy any power tool you should decide what you need it for. If you aren't going to do any major renovation work and the projects you intend to work on aren't too large, then the cheaper light-duty power tools may do just fine. On the other hand, if your tools are going to be put through the rigors of house renovation or some heavy-duty building, spend the extra money and get the appropriate tools. They are worth it.

AMPERES / HORSEPOWER	LIGHT DUTY	HEAVY DUTY
¼" DRILL	1.5 / ½	2.5 / ½
⅜" DRILL	2.5 / ¾	3.5 / ¾
7¼ CIRCULAR HAND SAW	7.5 / 1½	8.5 / 2
8¼" CIRCULAR HAND SAW	9.0 / 1¾	12.0 / 2
ROUTER	3.1 / ¾	8.0 / 1½
ORBITAL SANDER	1.5 / ½	3.0 / 1
BELT SANDER	6.0 / 1.5	10.5 / 2.0
SABER OR BAYONET SAW	2.3 / ¼	3.5 / ½

THE PLUMB BOB.

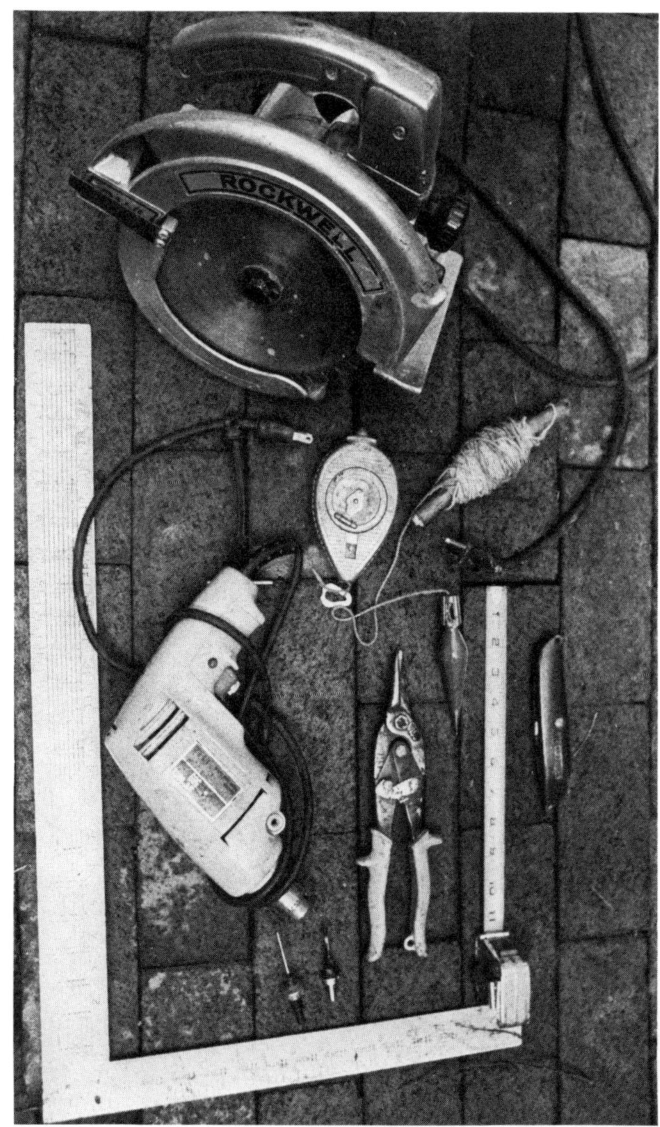

Your Tool Box

Since we now have it all together with our tools, let's get a place to stash 'em. My dad built a box for his tools and when he retired I stole it from him and it is still good today. Making a good box is so simple that it is well worth the effort.

Build it good and it will last you a lifetime.

The best thing to use for this box is ¾" and ½" plywood, but you can use packing crates if you want to. Although nails will work, screws and glue are the best. If you have any waterproof glue, dig it out for this one. If not, Elmer's is real good. 1¼" #8 wood screws (W.S.) will do just fine.

The base and ends should be made out of ¾" material, while ½" is just right for the sides and any internal dividers you may want to add (compartments for general tools and for special tools). If you own a router and really want to make the box super, then dado the corners. The handle can be made from an old broom or mop. If you are going to buy a dowel, then get one that feels comfortable in your hand.

The measurements for this box are the ones my dad used; but look at your tools and build your box accordingly. It seems that anything longer than 32" gets to be a little hard to manage. The ends should be about 10" high; the sides about 5½"; since the ends and bottom are the same width they can be cut from the same piece of wood.

The dividers which are put in the box are made for your own needs. My box, as you can see in the picture, is

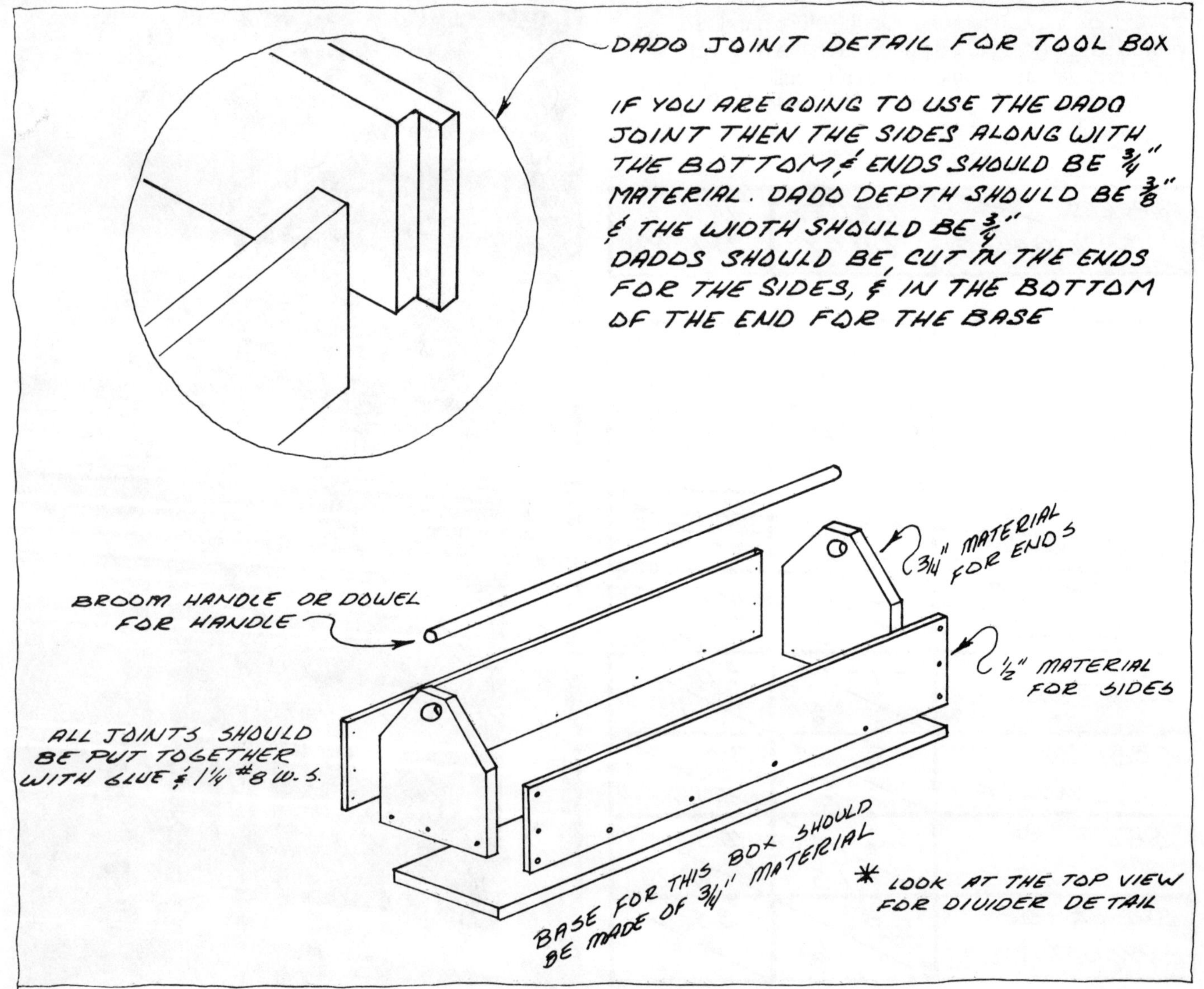

divided into three compartments: one for my saws, one for regular hand tools, and a third smaller section for my rulers, chalk lines, sharpening stone, etc. I usually keep those things I don't want bounced around in that third section. Along the center of the box I sectioned the center divider to store screwdrivers and chisels. The divider is made out of ½" stock (to keep the weight down). But since it is difficult to nail into an end of that width, I have used blocks to nail the divider to. Also use the block to nail the corners up. Nail both sides of a corner to the block for the thickness. The block should run the full length of the corner; ¾" x ¾" square is fine. The assembly of the box should be thought out in advance to make it easy to construct.

After you have collected your materials, including screws and glue, then you are ready. Cut all pieces to size, cut the angles in the ends, and drill the handle holes. When you drill the holes for the handle, make sure that there is enough material to hold so that the handle will not pull out of end. Next glue and screw the ends to the bottom. (Use a screw set in your drill to make the installation of screws easier and better.) Before you put the sides on you should make the dividers, since it is easier to work without the sides. Use blocks of wood for spacers. I used the ½" material, cut into strips, for the sectioned divider down the center of the tool box. When you have put in all the dividers that you need, then screw and glue the sides to the base and ends. Install the dowel handle. To fasten the handle to the box, nail through the dowel from the top of the end. Put in your tools, lay back and beam—and get ready to build a skylight.

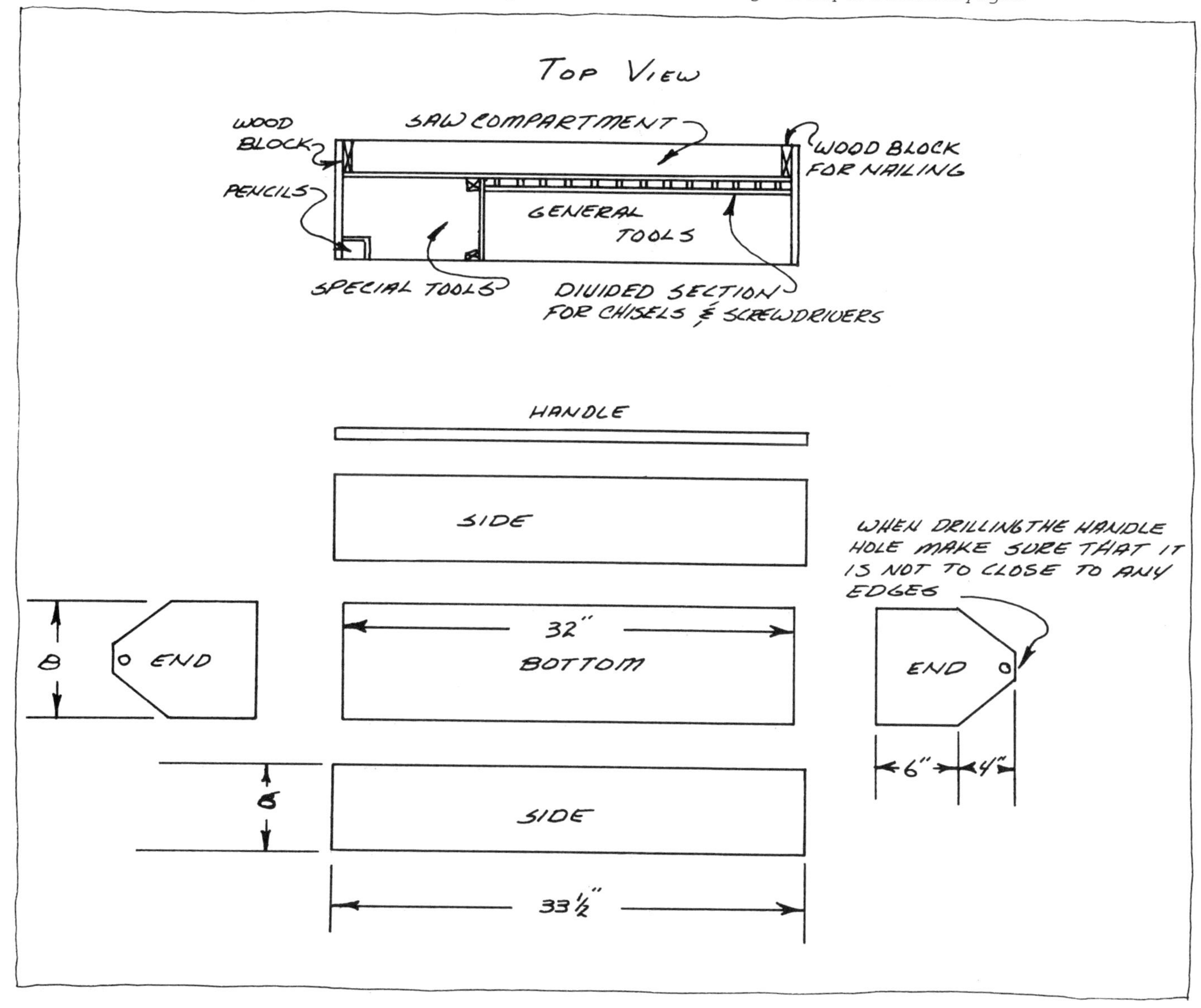

PLATFORM WORK AREA BUILT ON THE CEILING JOISTS.

Hints on Safety

Climbing on the roof, cutting holes, climbing ladders: these things can be very frightening and may keep you from building a skylight. But that need not be the case. If you learn a few roof rules and follow them, you will have a beautiful skylight and a job safely done.

The first rule should be followed carefully and given your full attention: *mark off the work area* on the ground where pedestrians or vehicles could be hit by a stray scrap. Use ropes or long pieces of wood to mark off the area. Remember both front and back.

The second rule should be in your head every step of the way: *watch where you are walking.* One of the first things you should do to prepare for your walk on the roof is to inspect it from underneath to make sure that there are no rotten roof boards which your foot could go through. Use a screwdriver as a probe to check for rotten wood. If you find soft wood, repair it soon and avoid walking on the area until it is repaired.

On the roof, you should always *wear rubber solid work boots.* If there is any dampness on the roof don't walk on it until it has dried fully. When you are walking on the roof, try to develop this little trick: *spot where you are going to put your feet* before you put them there. It may save a nasty spill or a terrifying slide down the roof.

When you are working on your skylight, try to *do as much as possible from the inside.* If you are working above the attic you can build a work platform on the ceiling rafters. And if you are working in an open, high-ceiling area within the house, you can build or rent a scaffold. Should you have to go over 10 feet high, I would recommend renting a scaffold. It is relatively inexpensive and makes the job much easier.

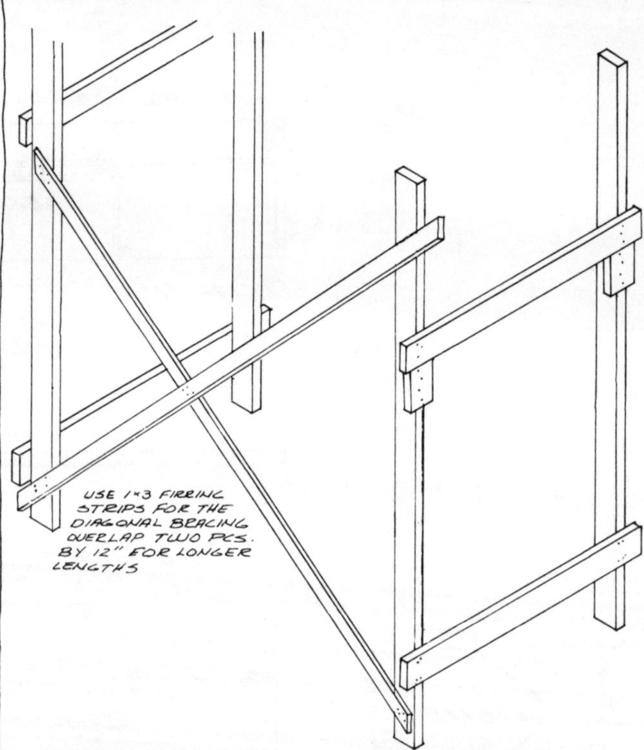

THIS SCAFFOLD SHOULD BE USED WHEN THERE IS NO OTHER PLACE TO ERECT A WORK PLATFORM. IF THE WORK HEIGHT IS OVER 8' A COMMERCIAL SCAFFOLD SHOULD BE USED FOR SAFETY REASON

2×4'S SHOULD BE USED FOR THE UPRIGHTS AND THE CROSS MEMBERS USE 12 PENNY COMMON NAILS

SOLID FOOTING IS CRITICAL WHEN WORKING WITH POWER TOOLS.

While you are working on the roof, *keep the work area very neat* for two reasons. Obviously, for general safety: if the work area is clean there is less of a chance that you will trip over something. The second reason for keeping the work area clean, especially free of nails and screws, is to avoid putting a nice hole in the roof's covering—usually tar paper and shingles—by stepping on a nail. A nail head can puncture your roof surface easily and potentially cause a leak.

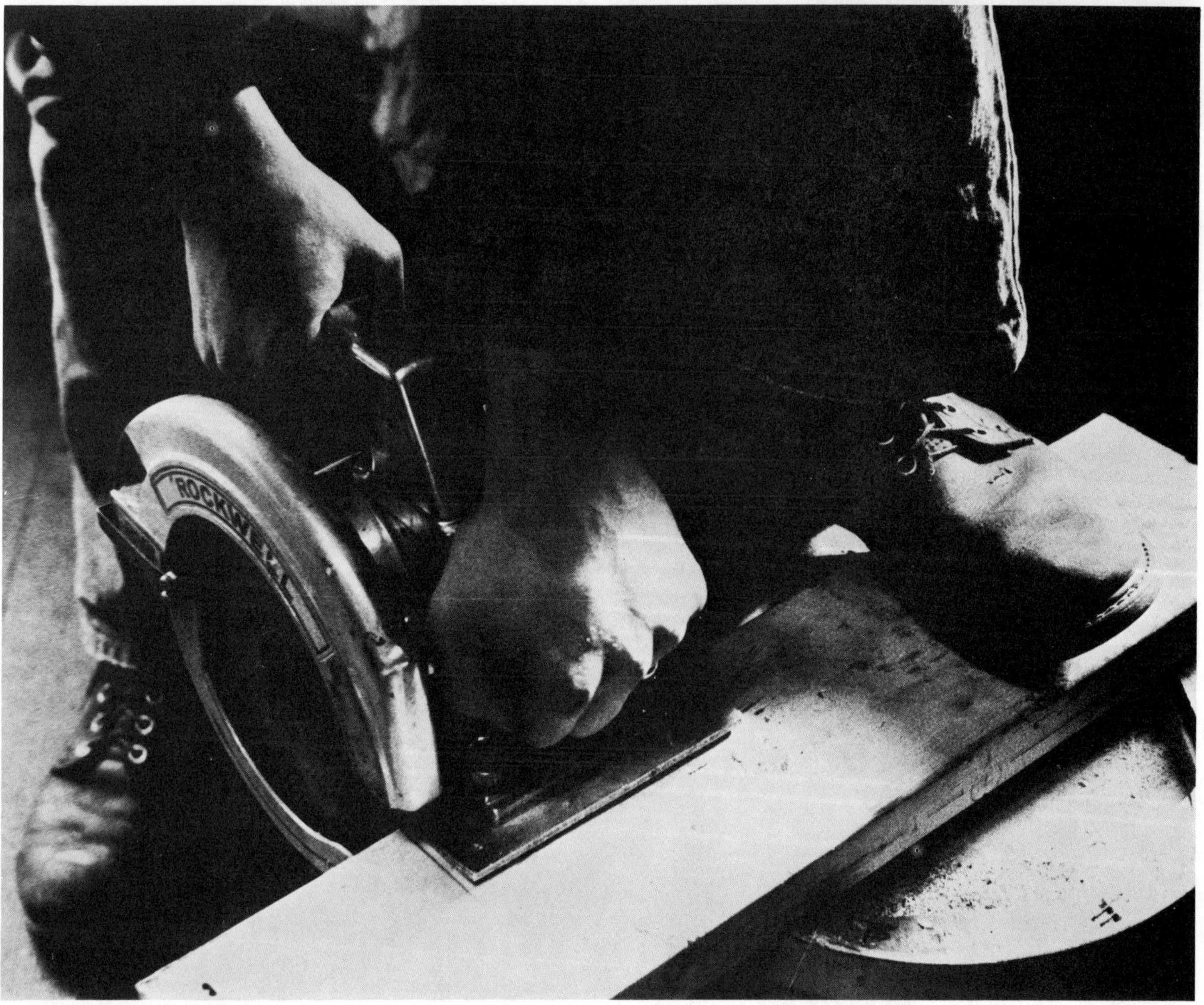

PUT THE ROOF CEMENT IN A PLACE WHERE IT WON'T ROLL OFF THE ROOF. THE CROTCH OF THE CHIMNEY IS A GOOD PLACE.

BECAUSE OF THE HARMFUL PARTICLES IN DRYWALL PASTE, PROTECTIVE EQUIPMENT SUCH AS GOGGLES AND MASK SHOULD BE WORN.

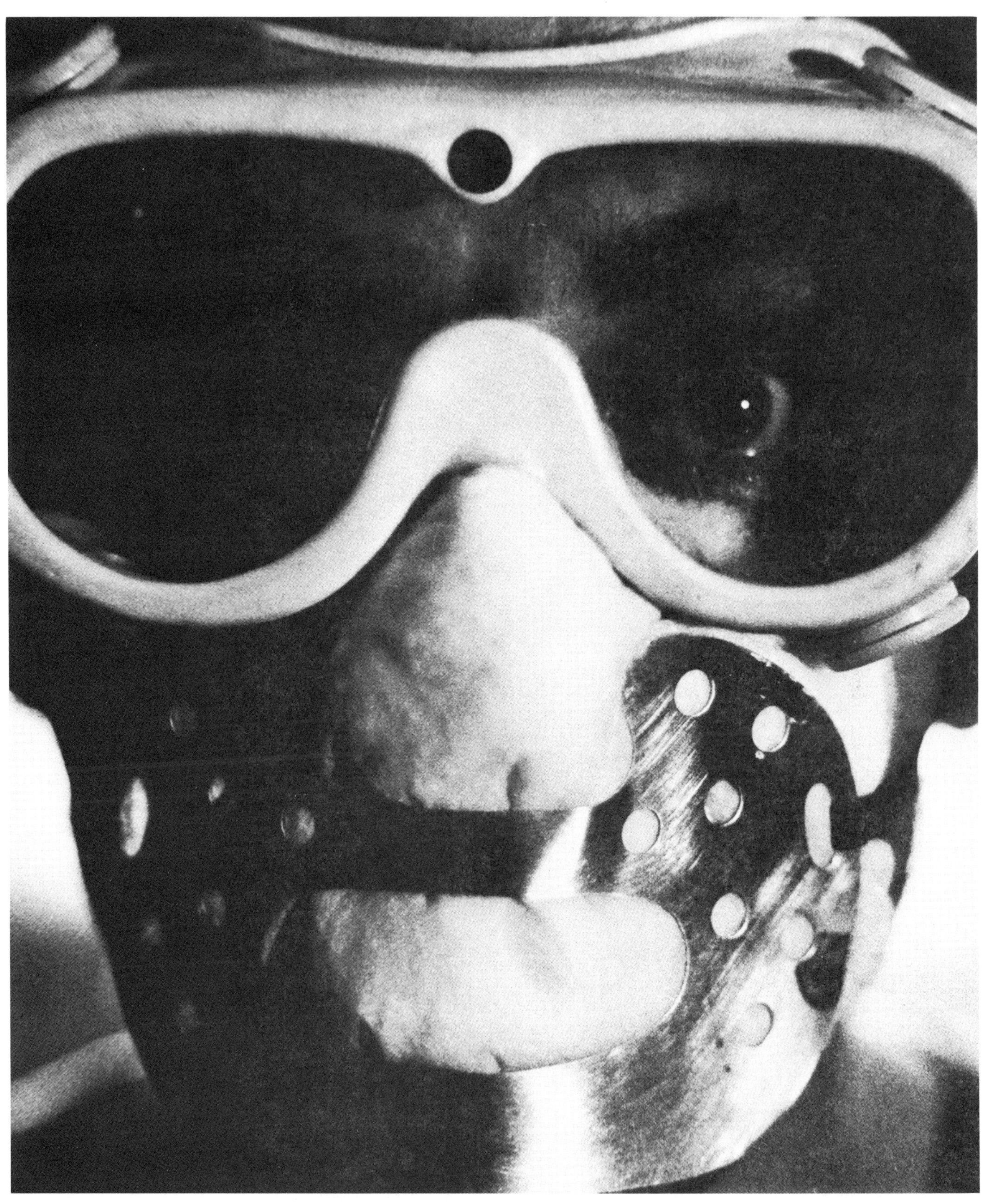

Curb Installation: The First Step in Construction

Materials

Here is a list of materials that you will need to have on hand in order to completely erect and install the curb.

> Make sure you have all the materials you will need to finish the skylight before you cut the hole in your roof so you will not have to stop in the middle of something to go to the hardware store.

2 x 6 material — This is what is used for building the curb itself.
Header material — The header material should be the same size stock as the ceiling rafter.
Aluminum flashing — This stock (obtainable in any hardware store) should be 9" wide; it is purchased by the foot.
Roof cement — Comes in both 1 and 5 gallon quantities. Although you won't need all 5 gallons, it is cheaper to buy it this way than by the one-gallon can.
Epoxy cement — Any epoxy cement works good here. PC-7, a two-part epoxy resin, is the one I used. It is used to seal the corners of the flashing.
12-Penny common nails — These are used for nailing the curb together, nailing curb to roof sheathing, and for nailing headers to rafters.
8-Penny common nails — Used for nailing curb to roof and bracing.
1" Galvanized roof nails — For nailing the roof material and for nailing the flashing to the top of the curb.

If you have the things on the list, then you are ready to begin work on your skylight.

Making the Curb

Regardless of which type of skylight you decide to use, either homemade or commercially manufactured, the first step in most cases is to build and install a *curb* for the skylight to sit on.

The curb itself is the first thing that has to be made. It is made of 2 x 6 material that should be free from all but the tightest knots. Try to get pieces which are not warped or bent. Pieces which have cracks at the end (this is known as checked lumber) should also be discarded.

To determine the correct measurements for the curb, look at the drawings in Figure 1. These drawings show how to get the right curb measurements in relationship to the skylight you have decided to use.

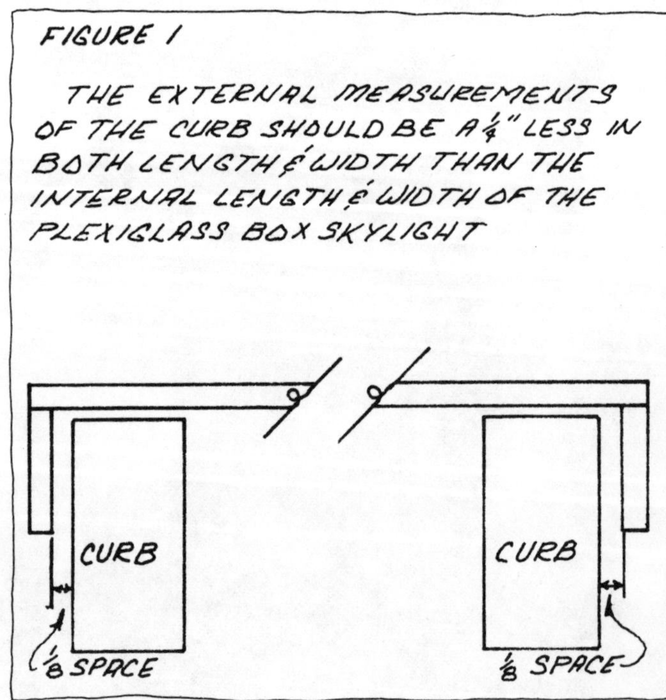

THE COMPLETED CURB WITH CORNER BRACES.

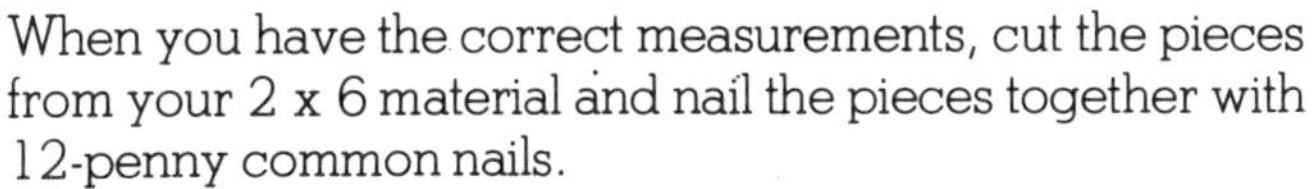

When you have the correct measurements, cut the pieces from your 2 x 6 material and nail the pieces together with 12-penny common nails.

> Let me add a word of caution here. When you are measuring and cutting anything, take your time and do it carefully. Try to remember this corny adage my dad used to tell me after I had made a bad cut, "Measure twice, cut once." Corny but true.

After the curb is nailed together, then the temporary corner braces are added, as shown in Figure 2. These are installed to keep the curb in square until it gets set on the roof. You can make a brace out of any piece of scrap, as it is removed when the curb is nailed in place. Nail it in with 6- or 8-penny common nails—but don't nail them in all the way, so you can pull them out easily.

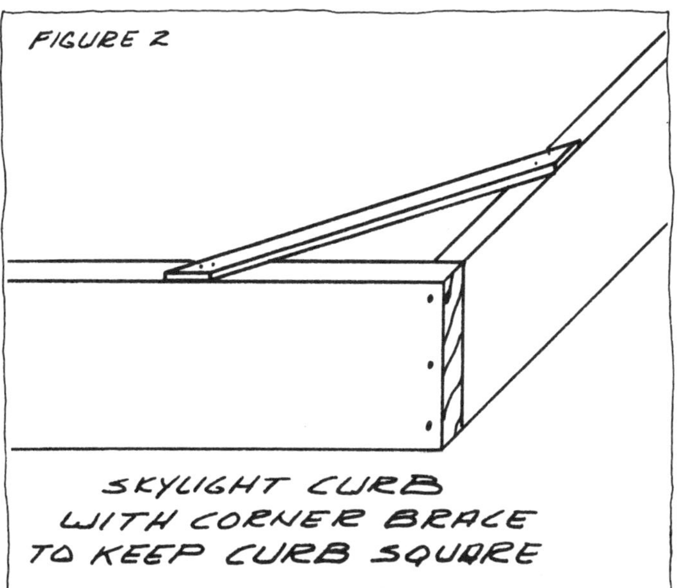

THE RAFTERS BEFORE CUTTING.

Cutting the Hole and Installing the Curb

The next step is to lay out the hole for cutting. Check to make sure that you have the right measurements: *the internal measurements of the curb should have the same dimensions as that of the opening.* After checking the

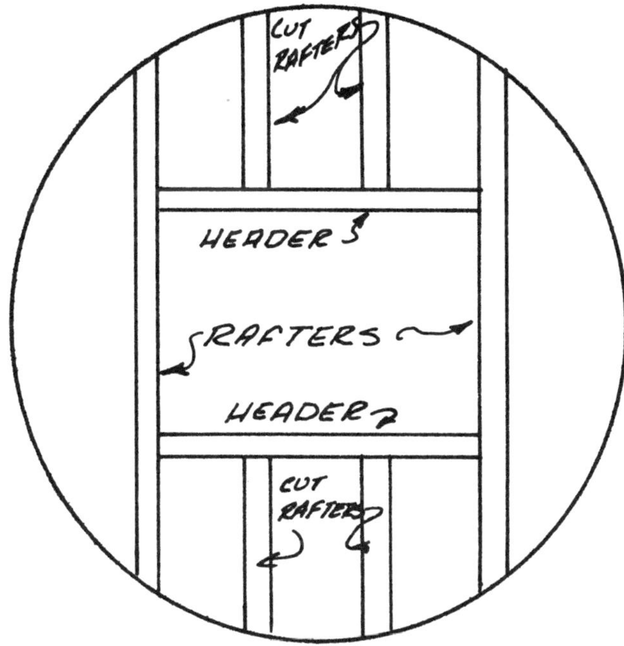

measurements, cut any rafters which need cutting. Remember, when you are cutting the rafters they must be cut back 1-5/8" on each side to allow for the nailing of the headers, as in Figure 3. Do not cut into the sheathing just yet—cut and remove just the rafters.

LAYING OUT THE SKYLIGHT HOLE.

THE SHEATHING BEING CUT FROM BELOW.

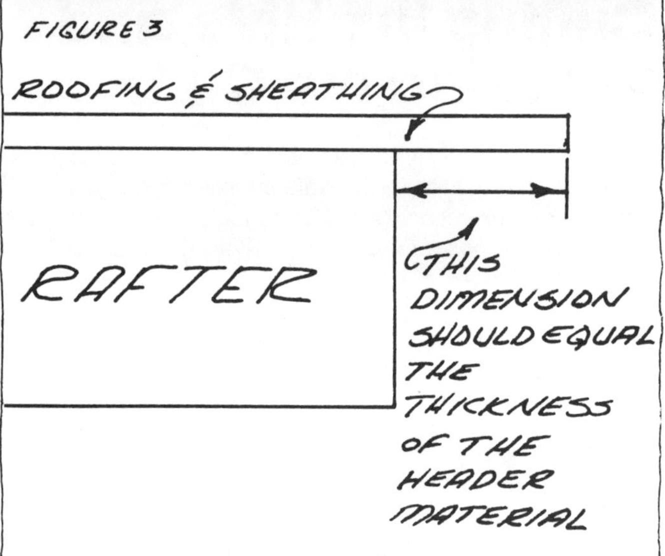

There are two ways to cut the sheathing and roofing material. One way is to,cut out everything from the inside. The advantage of doing it this way is to eliminate some of the climbing on the roof. The problem with this, however, is that the cuts must be made over your head—which can be very difficult with a power saw. To cut the skylight from the inside, just set the saw to the thickness of the sheathing and cut along the line you laid out before. After the sheathing is cut, then cut the roofing material with a mat knife for shingle roofs or a flooring chisel for tin or composite roofs.

If you don't want to deal with a power saw over your head, there is another, and, I think, easier way to get this part done. Drive a nail up through the roof at each corner of the opening you are going to cut. After this is done, go out on the roof with the curb, power saw, flooring chisels, hammer, and chalk line. If you have a tin roof, then take up the tin snips too. When you have located the four corners, take the curb and put it over the four corner nails. Trace a line around the *outside* of the curb. This line indicates how far the roof material is cut back. Use the flooring chisel and mat knife to cut the roofing material away. After the roofing material is cleared away, lay the curb back down on the now bare sheathing and trace a line around the *inside* of the curb. Remove the curb and cut along the line you have drawn to open the roof. After you are done cutting, the best thing to do is clean up the area. Check the roof especially carefully for loose nails; they can cause you to loose your footing quite easily. After you are done cleaning, take a little break — I'm sure if you have any friends helping, they won't mind a bit.

The cutting and setting of the *flashing* is the next step. The flashing material is made of aluminum and can be purchased at most hardware stores and lumber yards. The flashing can be cut with tin snips or a mat knife or even a pair of heavy scissors.

Study the patterns in Figure 4. Take your time when cutting the flashing so it comes out correctly. This will eliminate the chance of leakage. The overlap at each end of the flashing is 3 inches and the flap that slides under the roof is also 3 inches. Use the curb to help you form the flashing. After you have all the flashings formed, go up on the roof and make sure the roof material is free from the sheathing for at least 4 inches around. Make sure there are no nails to stop you from sliding the flashing in. Take a look at the drawing in Figure 5.

Take the flashing up on the roof along with the five gallons of roof cement. Make sure you bring a pair of

CUTTING THE ROOF MATERIAL WITH A FLOORING CHISEL.

REMOVING THE OLD TIN ROOF.

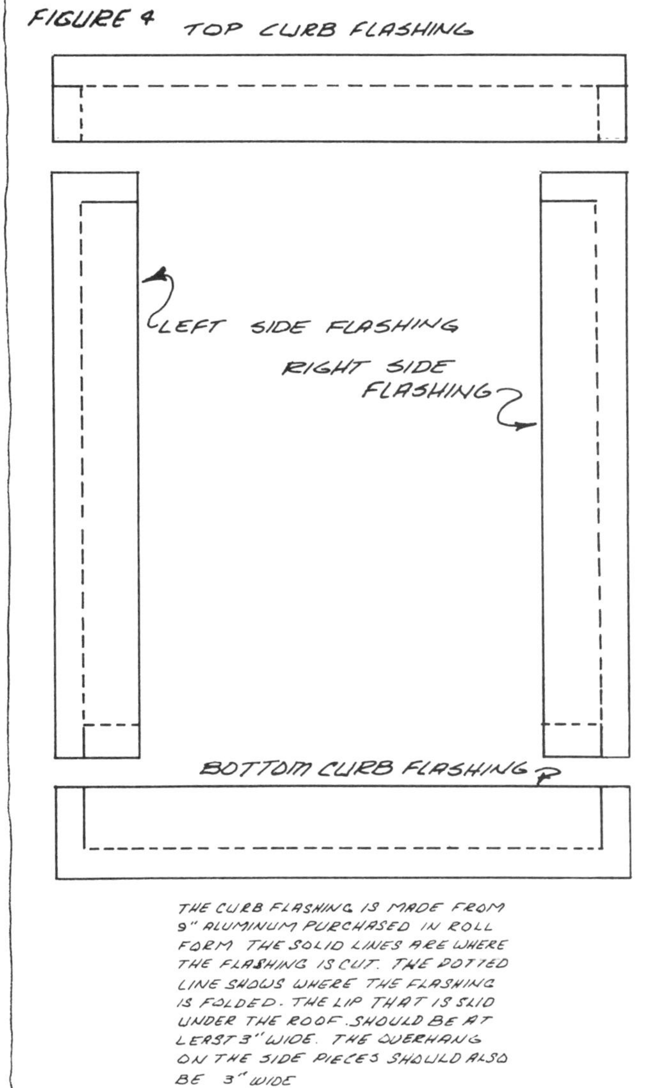

old gloves, some paint thinner, and rags—this job has a tendency to get a little messy. Coat only one side of the skylight at a time to keep the mess down.

After you have put down a good layer of the roof cement, take the bottom piece of flashing and slide it under the roof incline. Next do the left and right sides. First apply a coat of cement and then slide in the flashing. The top is put in last (Fig. 6). Bind all the flashings back so they

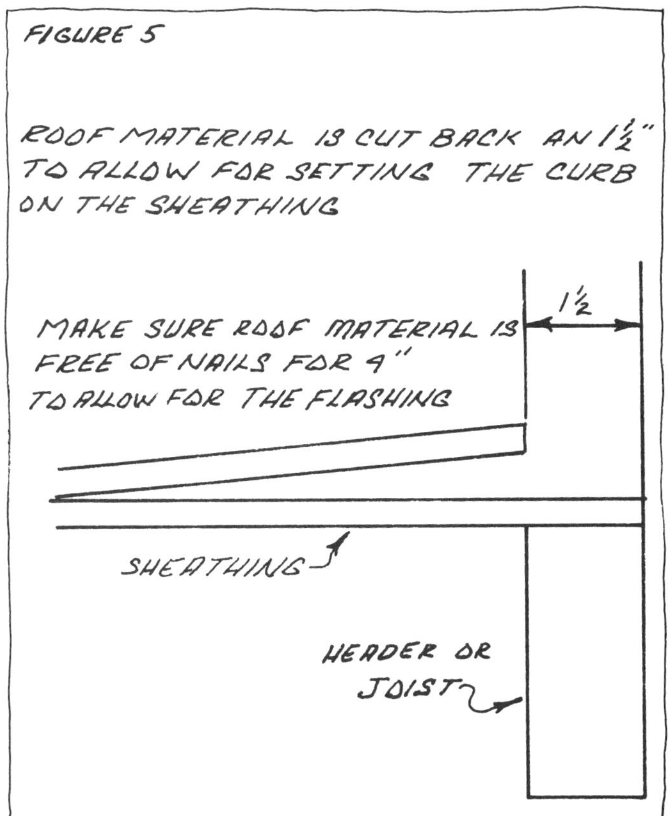

TRIMMING THE TIN. GLOVES SHOULD BE WORN TO PROTECT THE HANDS.

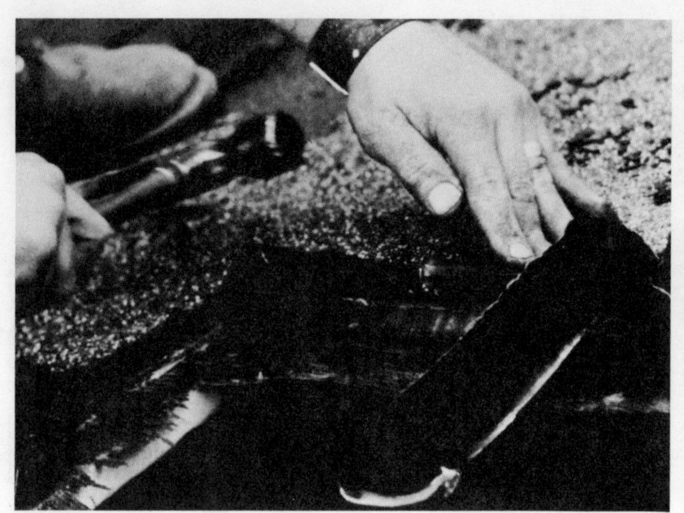

NOTICE THE ROOF MATERIAL CUT BACK FROM THE SHEATHING TO ALLOW FOR THE CURB.

FIGURE 6

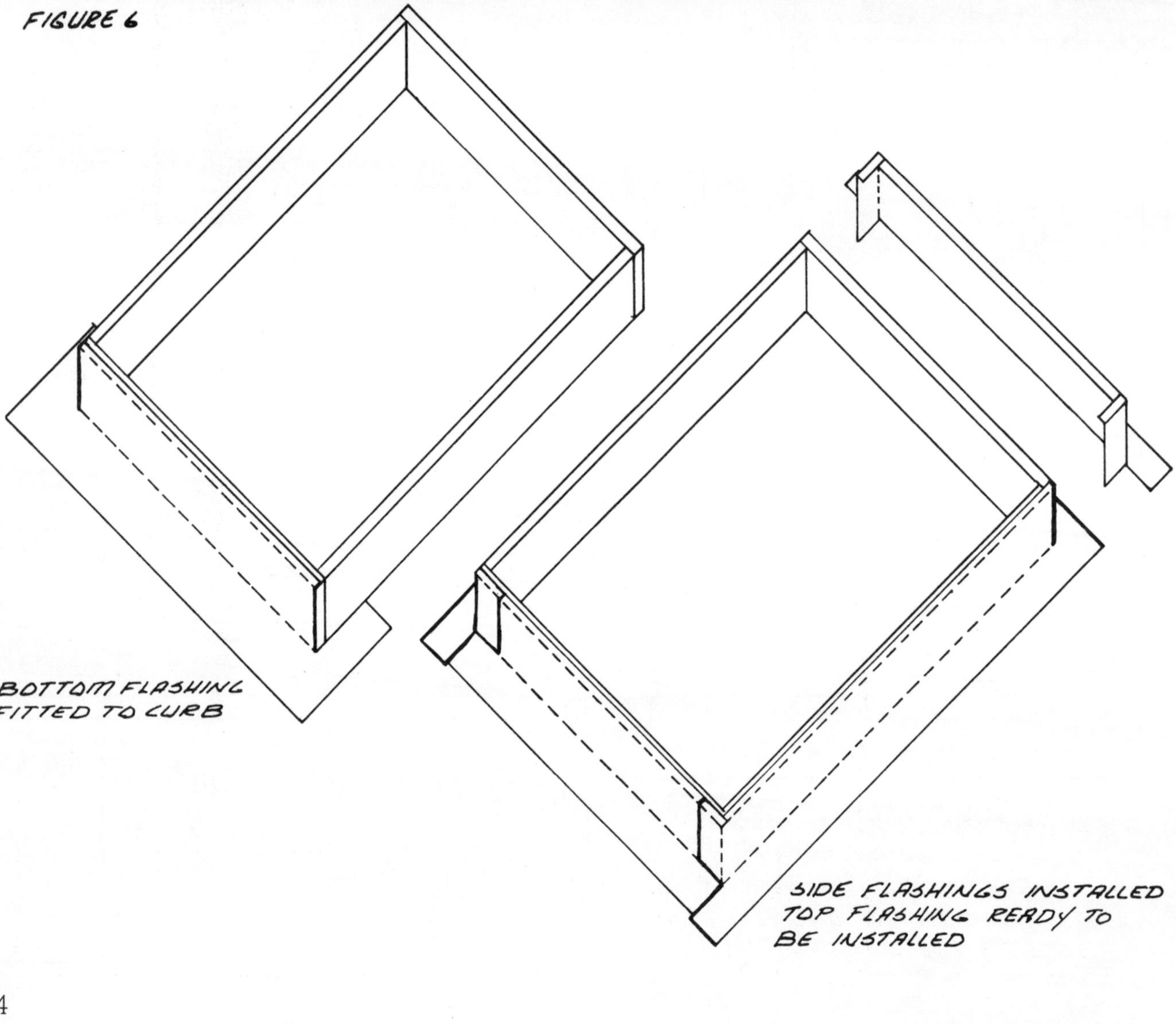

BOTTOM FLASHING FITTED TO CURB

SIDE FLASHINGS INSTALLED TOP FLASHING READY TO BE INSTALLED

THE ROOF MATERIAL IS PRIED UP TO ALLOW FOR THE INSTALLATION OF THE FLASHING.

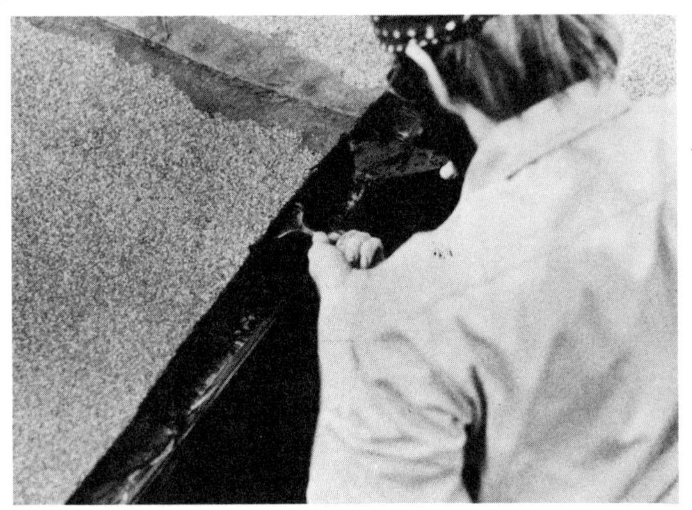

APPLYING A COATING OF ROOF CEMENT TO THE SHEATHING IN PREPARATION FOR THE FLASHING.

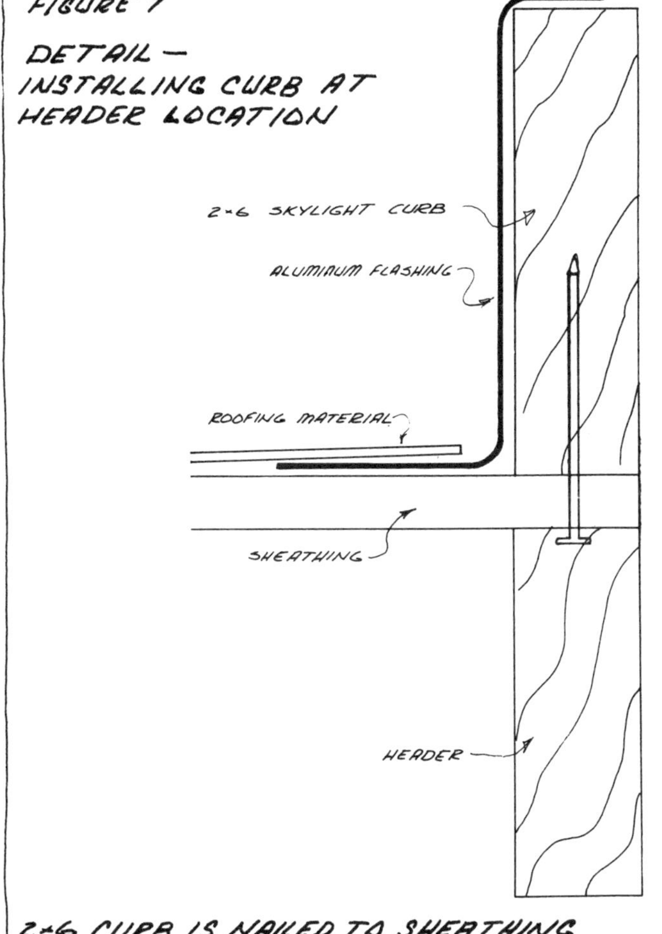

will not get in the way of setting the curb—which is the next step. The curb should sit down on the sheathing and be nailed around the edges of the sheathing with 12-penny common nails (Fig. 7). Where there are rafters, toe-nail the curb to them with 8-penny common (Fig. 8). Next bend up the flashing and nail around the

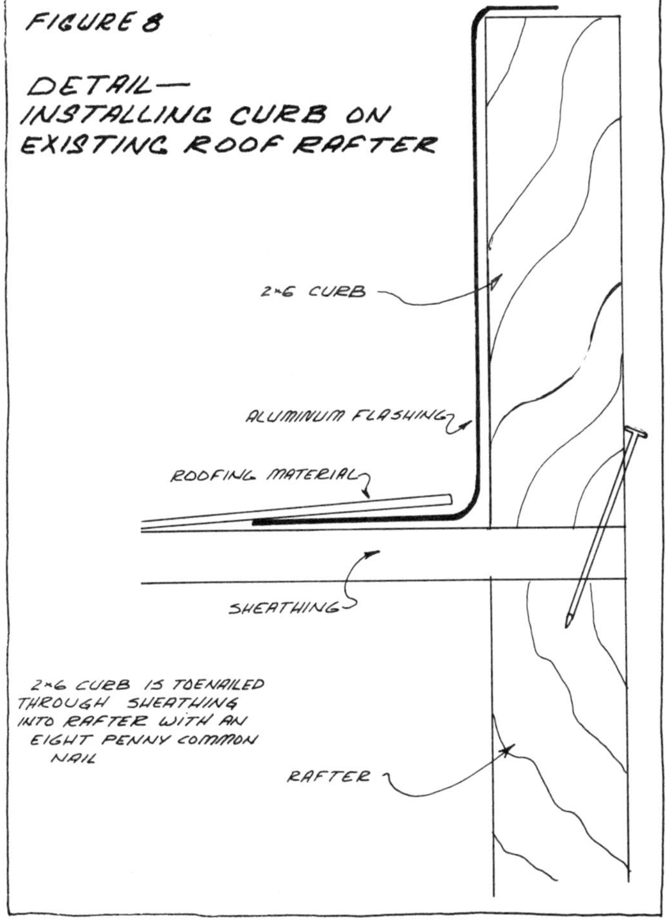

THE FLASHING IN PLACE AND THE CURB BEING TOE-NAILED TO THE SHEATHING.

top of the curb with 1" galvanized nails. Cover the corners with a coat of epoxy (PC-7 is excellent here). Please follow the directions on the can. Cover the edge of the roof material with a coat of roof cement as in Figure 9. Clean up the area, your tools, and yourself—and that just about does it for installing the curb.

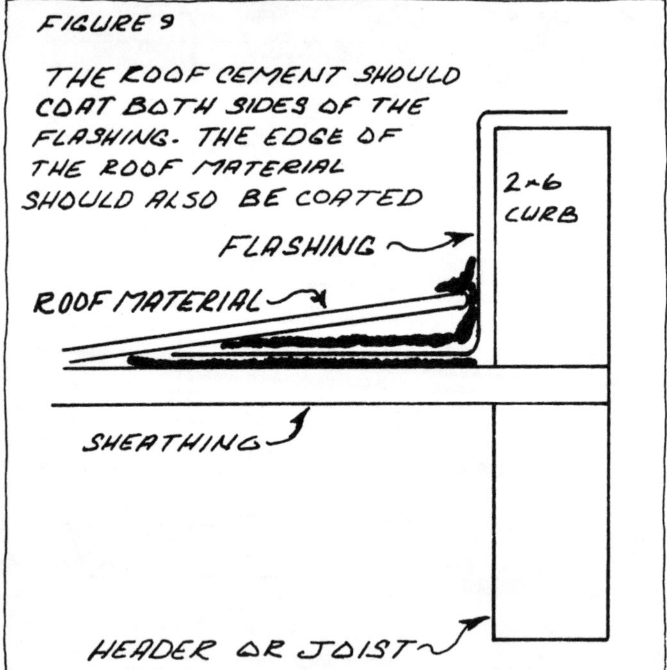

The only thing left is to nail the roof material back in place. Make sure that you do not put any nails into the flashing itself. Cover all exposed nails with a coat of roof cement. Then, nail in the headers. Use 12-penny common nails. The headers should be of the same material as the rafters. This finishes the installation of the curb.

THE COMPLETED CURB.

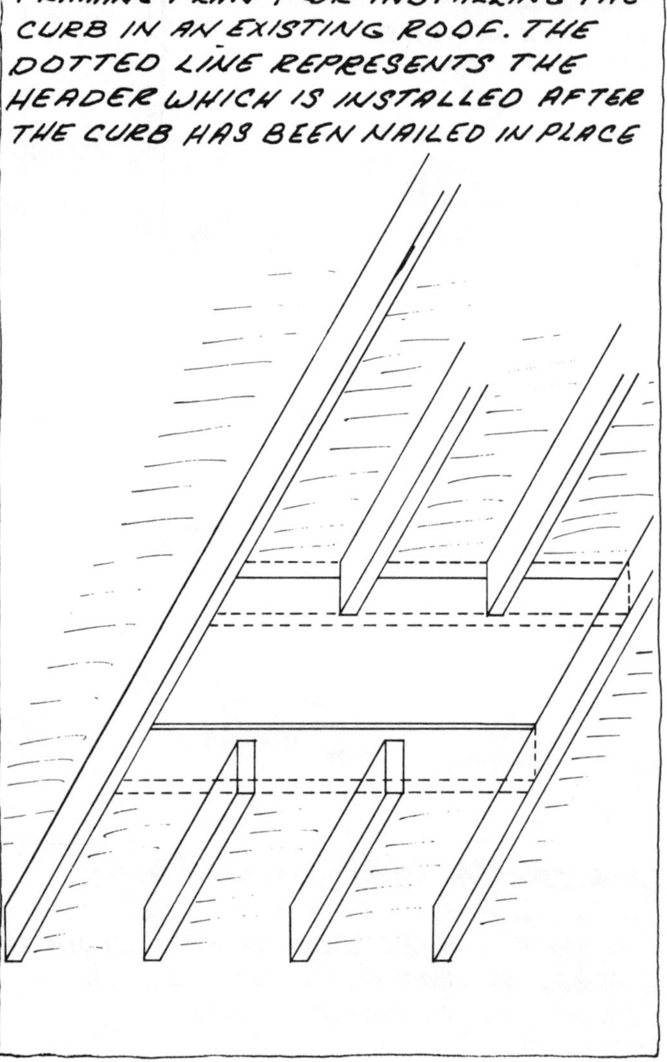

The Plexiglas, or Plate Glass, Skylight

There are many different kinds of translucent materials that can be put on the curb. For the person building his own, the Plexiglas or plate glass skylight seems to work very well—and it seems to me to be one of the easier skylights to build. As with any construction work, your craftsmanship must be of the highest quality in order to get the best skylight possible and avoid leakage.

After the curb has been erected, check to see if all the top edges of the curb are even. If not, sand or plane them flush. After this is done the next step is to nail a spacer, called a batten, around the outside edge of the curb. This will hold the aluminum angle at the proper space above the plate to allow for the silicone caulk. The batten strip should be about ½" to ¾" wide, depending on the size of the skylight material; the gap between the batten and the material should be no less than ¼". The thickness of the batten should be twice the thickness of the material if you are using Plexiglas. If your material is plate glass, allowing a gap of 1/8" is sufficient. Use 4-penny finish nails to nail the batten around the edge.

The next step in building this skylight is to prepare the *aluminum angle.* With the aid of the silicone caulk, the angle is used to hold down the Plexiglas or plate glass. Cut the aluminum angle to size by using the curb for your measurements. After the aluminum angle is cut, you must drill some holes in the angle for the hold-down screws. The holes should be countersunk to allow for the seating of the faucet washer gasket (see Fig. 10). These holes should be about 12" on center. A 1" #8 wood screw works very well here.

When the aluminum angle is prepared and the curb is ready to receive the skylight material, plane or sand smooth and have the batten nailed in place; then lay a bead of silicone caulk around the curb or ½" from the inside edge of the curb. When you cut the top off the caulk tube, don't cut a hole over 5/16" wide; ¼" is the best. After the caulk is layed out, peel one side of the protective paper off the material, set it over the curb, and press it into place on the silicone caulk. Make sure it is evenly spaced between the battens.

> A little trick: if you are handling a large sheet of Plexiglas, lay some thin pieces of wood across the skylight so they rest on the battens; then you can lay the material down on these and position it before setting it in place.

Once the Plexiglas is ready to be set in place, just slide out the wooden strips one by one and let the skylight material settle into the silicone caulk on the curb. When this is done, take the paper off the remaining sides and lay down another bead of caulk on the skylight material. Make sure that the bead of caulk is higher than the batten so that when you set down the aluminum angle it will meet with the silicone to form the seal. After the aluminum is set and screwed into place, go around and coat all seams of the aluminum angle with a layer of the silicone caulk. Cover all the exposed screw heads with the silicone also. Check the flashing and the roof material, and with a layer of roof cement coat any areas which may seem like problem areas. When you have cleaned up all the scrap and loose nails, you are ready to go down and enjoy your skylight.

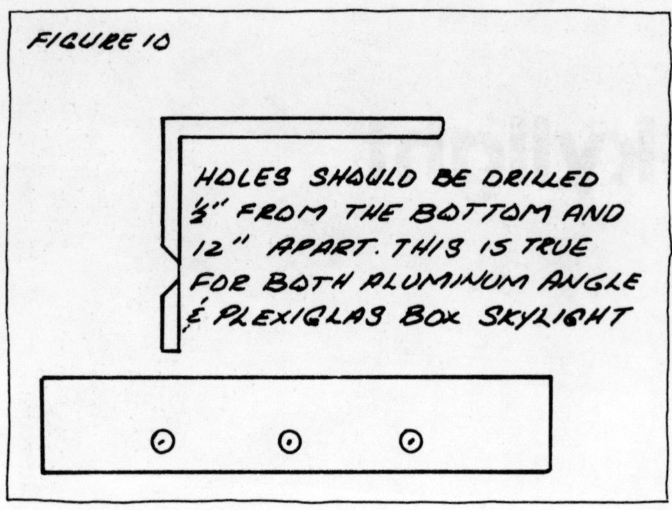
FIGURE 10
HOLES SHOULD BE DRILLED
½" FROM THE BOTTOM AND
12" APART. THIS IS TRUE
FOR BOTH ALUMINUM ANGLE
& PLEXIGLAS BOX SKYLIGHT

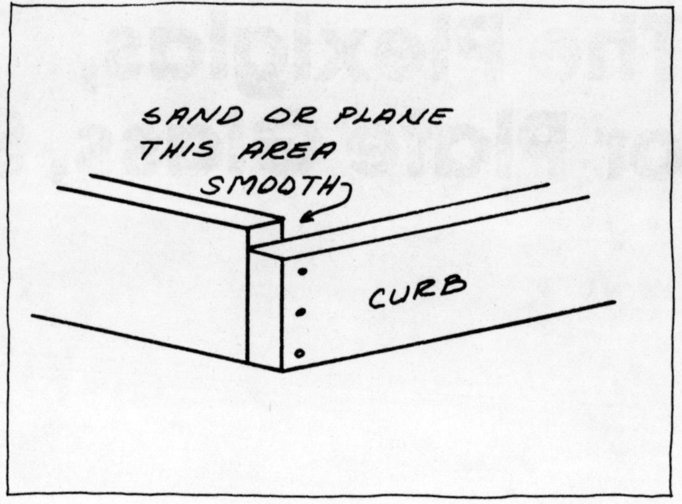
SAND OR PLANE
THIS AREA
SMOOTH
CURB

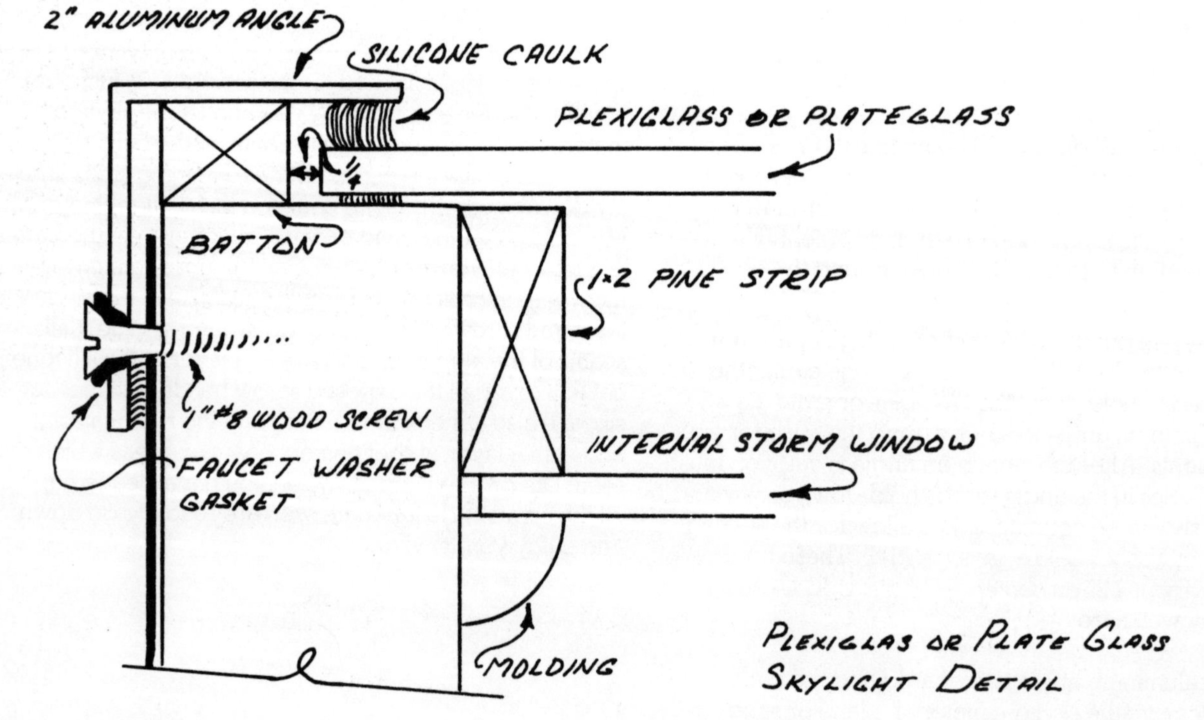
SOME GENERAL NOTES ON THE PLEXIGLAS
OR PLATE GLASS SKYLIGHT. THE BATTON
SHOULD BE A ¼" HIGHER THAN THE MAT-
ERIAL USED FOR THE SKYLIGHT.
THE 1" BRASS WOOD SCREW IS SET IN A
RUBBER FAUCET WASHER AND THEN
DABBED WITH A BIT OF SILICONE
CAULK
2" ALUMINUM ANGLE
SILICONE CAULK
PLEXIGLASS OR PLATEGLASS
¼
BATTON
1×2 PINE STRIP
1" #8 WOOD SCREW
FAUCET WASHER
GASKET
INTERNAL STORM WINDOW
MOLDING
PLEXIGLAS OR PLATE GLASS
SKYLIGHT DETAIL

MEASURING THE CURB.

CUTTING THE BATTEN

NAILING THE BATTEN.

CAULKING THE CURB.

MARKING THE ALUMINUM ANGLE FOR CUTTING.

INSTALLING THE ALUMINUM ANGLE AFTER THE SECOND COAT OF CAULK.

SCREWING IN THE ALUMINUM ANGLE.

WOOD SCREW WITH THE FAUCET WASHER GASKET FOR WATERPROOF ASSEMBLY.

FINAL COAT ON ALL SUSPECT AREAS.

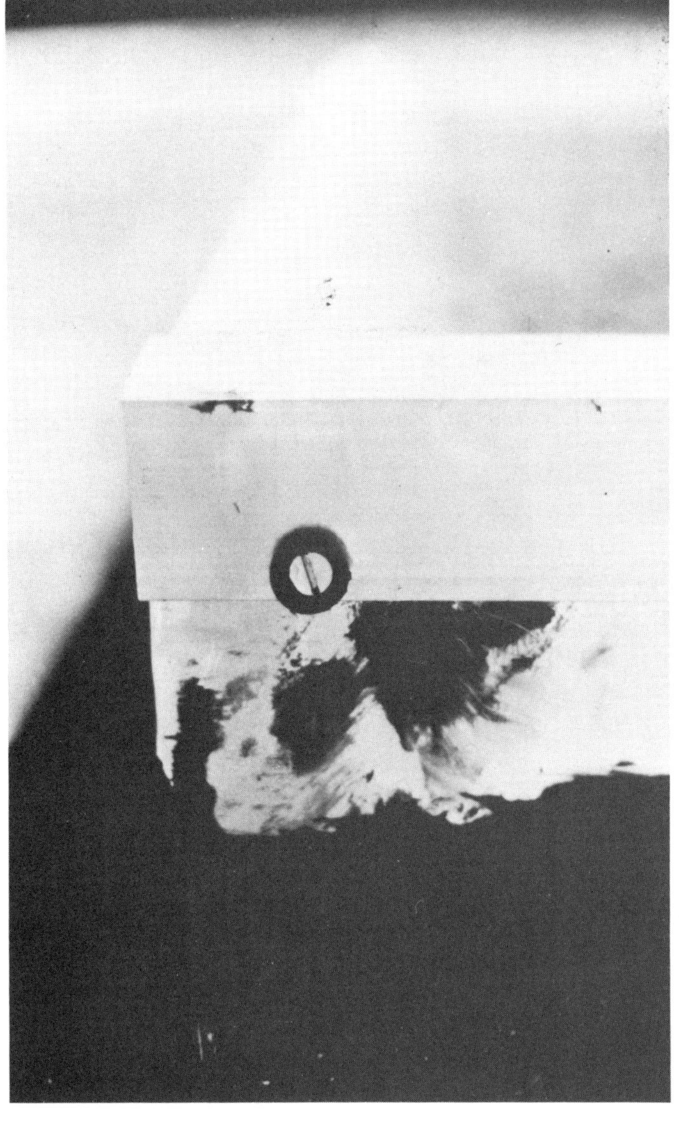

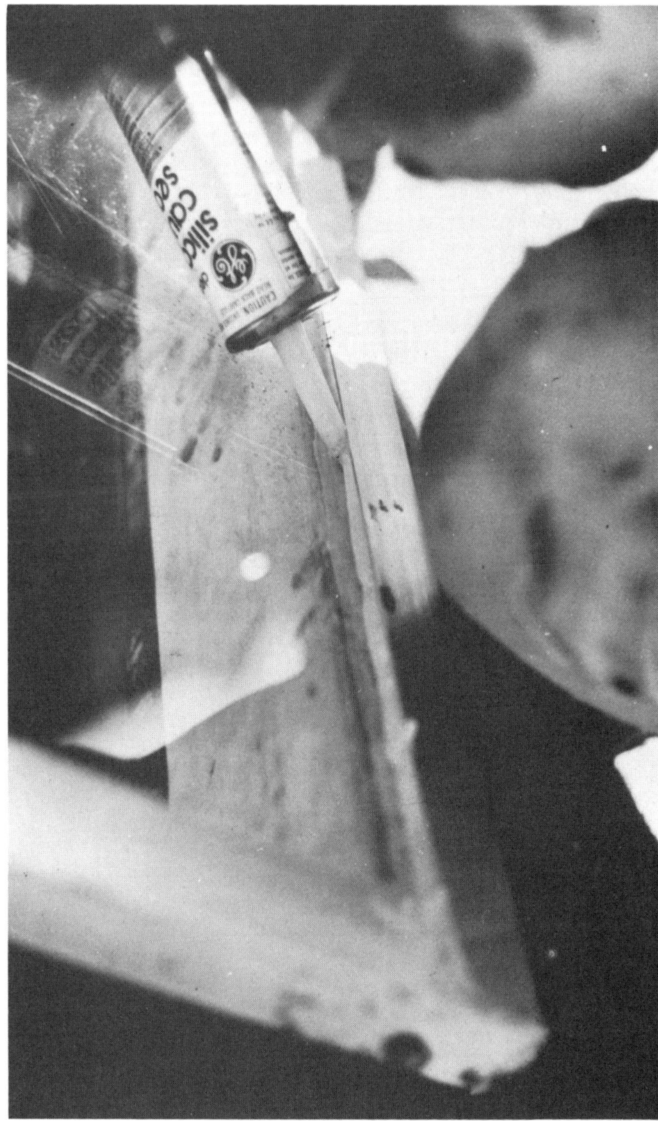

The Plexiglas Box Skylight

The Plexiglas box skylight represents a different solution to the problem of attaching the skylight material (whatever it may be) to the curb. In using the box skylight, you have eliminated the need for the batten strip around the edge of the curb. Also eliminated is the need for any aluminum angle. But you do have to deal with cutting and gluing the Plexiglas—which some people may prefer, rather than working with the batten and aluminum angle.

NOTE: If you are going to build this skylight please read very carefully the following section on "Tips on Working with Plexiglas." Also, try to pick up a booklet called "Do it Yourself with Plexiglas" from your Plexiglas dealer.

To get the proper measurements for the box skylight and to allow for the expansion and contraction of the Plexiglas you must measure the length and width of the curb. To these measurements add ¼" for each 4' of skylight material and add ½" for the two sides (¼" per side). These new measurements are the measurements of the top for the box skylight. Next you will need four strips to be used for the sides.

When you have these 4" strips cut to the proper length, the next step is to drill the holes for the hold-down screws. These holes should be countersunk to allow for the faucet washer gasket, as shown in Figure 10. These holes should be about 2" in from the end, about 1½" from the bottom, and they should be spaced about 12" on center (i.e.,

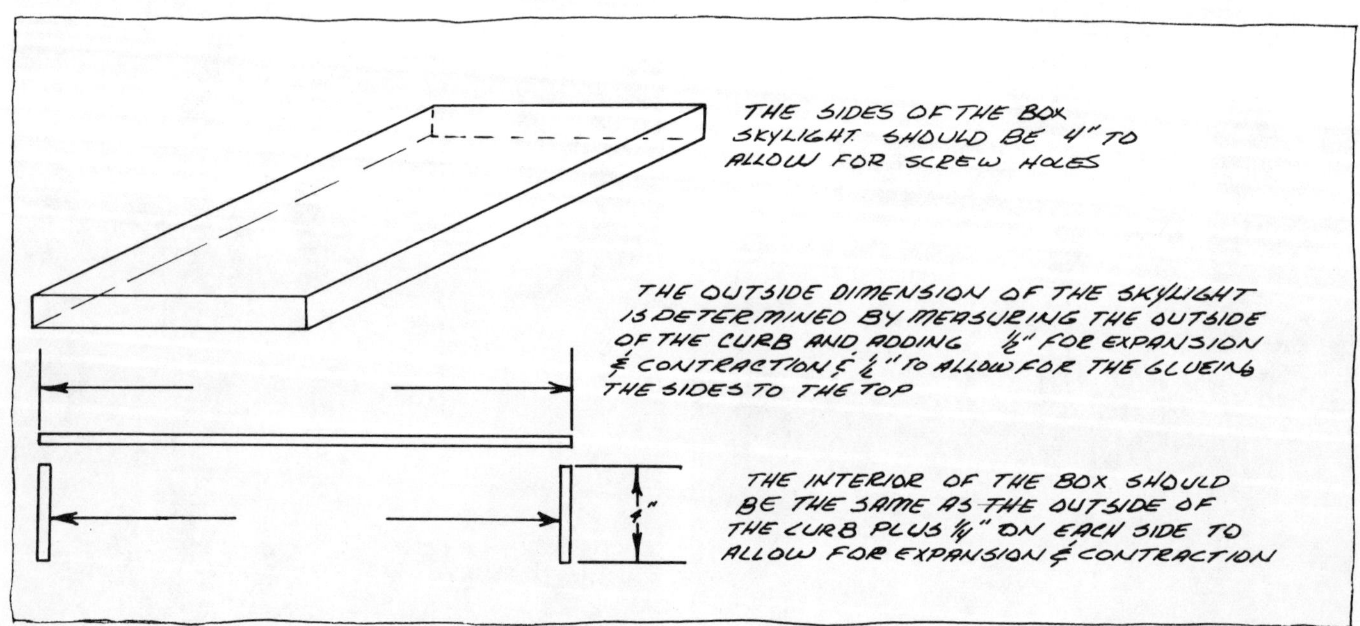

12" from the center of one hole to the center of the next). When you are drilling the holes, use a block of wood to back up the work in order to minimize the chance that the plastic might crack. Once the holes have been drilled, then glue the side strips to the skylight top, following the directions in the "Tips on Plexiglas" section.

After the skylight box frame has been made, lay a bead of silicone caulk around the outside top edge of the curb and set the Plexiglas box over the curb. When you set the Plexiglas box on the curb, slightly wiggle it around in order to help spread the silicone caulk a little bit. Fasten the box to the curb by installing wood screws and faucet washers (as a gasket) through the sides of the Plexiglas box and into the side of the curb. Pull off the remaining paper and enjoy the new found light.

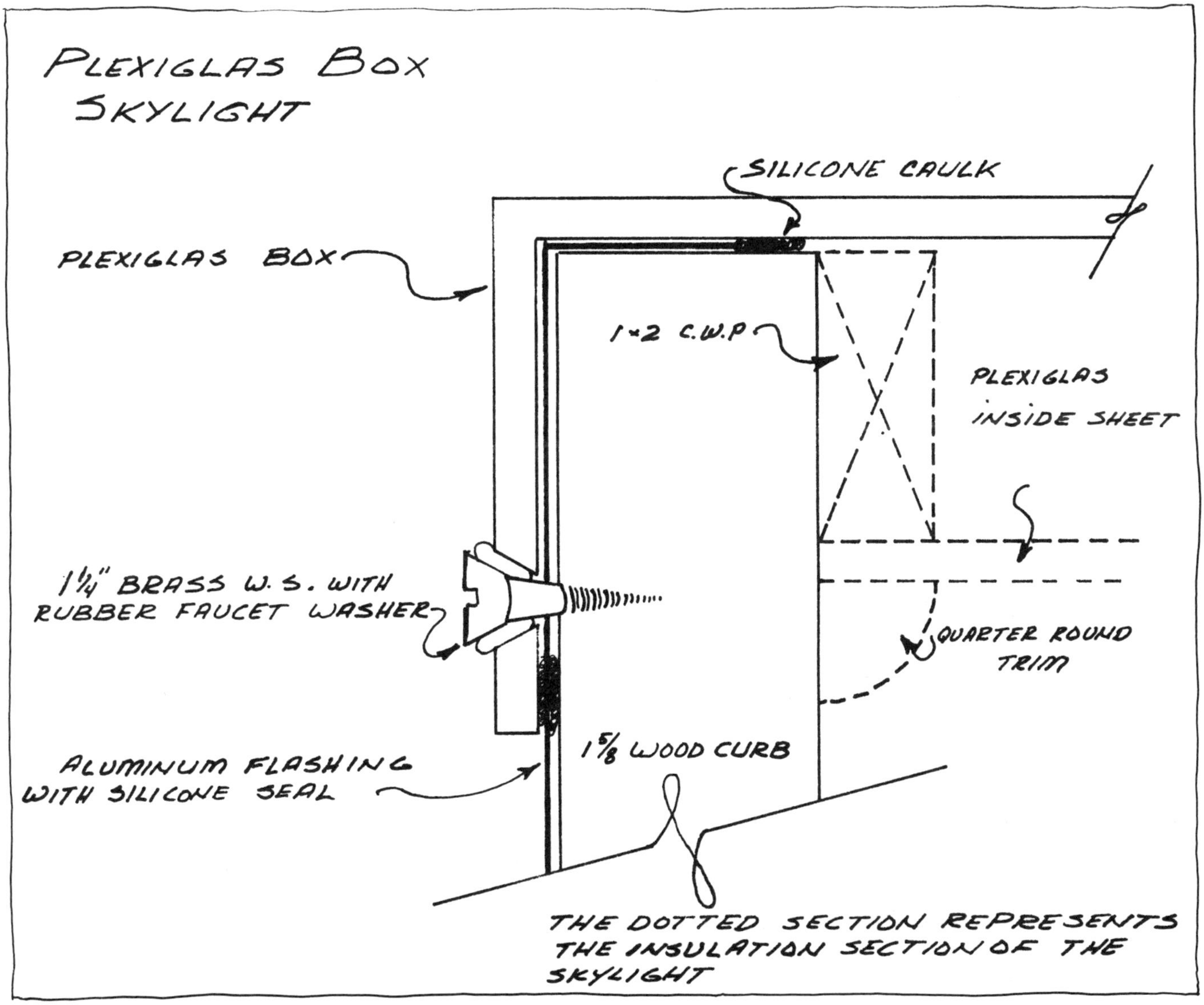

PLEXIGLAS BOX SKYLIGHT

SILICONE CAULK

PLEXIGLAS BOX

1×2 C.W.P

PLEXIGLAS INSIDE SHEET

1¼" BRASS W. S. WITH RUBBER FAUCET WASHER

QUARTER ROUND TRIM

ALUMINUM FLASHING WITH SILICONE SEAL

1⅝ WOOD CURB

THE DOTTED SECTION REPRESENTS THE INSULATION SECTION OF THE SKYLIGHT

Tips on Working with Plexiglas

You need actually to work on a piece of Plexiglas only for the box-style skylight. Cutting, gluing, and drilling Plexiglas are not hard to do at all—it just takes a little getting used to.

First, let's deal with *cutting.* If you are going to cut the Plexiglas with a sabre saw or scroll saw, use a saw blade with 14 teeth per inch when cutting the ¼″ Plexiglas. When you want to cut a straight line, clamp a straightedge to the sheet as a guide. Do the same for circular hand saws. The proper saw blade for circular hand and table saws is a crosscut blade of the type recommended for finish cuts on veneer plywood. Do not remove the paper when you cut the Plexiglas and don't force the feed of material. You have to hold the material firmly when you are cutting.

SCRAPING THE EDGE OF THE PLEXIGLAS WITH THE BACK OF A HACKSAW BLADE FOR A GOOD GLUE JOINT.

CUTTING THE PLEXIGLAS, USING A STRAIGHT PIECE OF WOOD AS A GUIDE.

When *drilling* Plexiglas, always back up the material with a block of wood and make sure that the drill bit you are using is sharp. With an electric drill, I recommend that you use Hanson's Special Purpose High Speed Bits; but if you are very careful, any good sharp bit will do. For small holes, like the ones we are using for the screw and faucet washer gasket, the highest speed is the best. When your bit is about to completely penetrate the material you are drilling, slow down to avoid chipping. After the hole is drilled go back with your screw, and set and countersink for the faucet washer. When drilling holes in the side of the Plexiglas box skylight, you have to consider somewhat the expansion and contraction of the Plexiglas. A rule of thumb is to drill holes 1/16" oversize for every foot on the side. Make sure there is at least a ¼" of material from the circumference of the hole to edge of the piece.

While there are various ways to finish the edges for the sake of appearance, we don't need to get into that here. If you are interested in finding out more, go to your Plexiglas dealer and ask for the free booklets that tell you about working with Plexiglas. Our main concern is to *prepare the edge for gluing.* To do this, first scrape the edge with the back of a hacksaw blade or sand it down with some 80-grit production paper. After the saw marks are gone, change over to 150-grit wet or dry paper to give you a clean satin finish. Be sure *not* to round the edges in the sanding operations. Now you are ready to start the gluing operation.

There are two types of cement for the Plexiglas: capillary cement and thickened cements. *Capillary cements* can only be used with Plexiglas G (general purpose). After the edge has been sanded satin-smooth, pull the paper back slightly and tape the two pieces together with masking tape. It is best to apply the capillary cement with a needle. (You can purchase an applicator from the place where you got the acrylic sheet.) The joint should be kept in the horizontal position to allow the cement to flow properly. *Thickened cement* is applied differently. After you have checked for a proper fit, apply a small bead of cement to one piece, then join the two pieces together and clamp until set.

> Both of these cements should be handled with great care. Avoid the vapors as much as possible.

TAPING THE PIECES OF THE PLEXIGLAS BOX SKYLIGHT BEFORE GLUING.

DRILLING THE PLEXIGLAS FOR THE FAUCET WASHER GASKET. NOTICE THE USE OF THE WOOD BLOCK UNDER THE PLEXIGLAS TO AVOID CHIPPING.

Checking for Leaks

Once your skylight is installed it should be checked for leaks before you remove all your work materials from the job site. If you have a hose that reaches to the roof, using it is an excellent way to water-test your workmanship. (If you have a problem getting a water supply onto the roof, wait until it rains.) Do this testing before the shaft walls are built so that you can see clearly if there are any leaks.

Once you have your water supply up on the roof, douse the whole skylight completely with water. Pay special attention to the location where the aluminum flashing slides under the roof covering material; this will probably be the place where leaks will develop. Have a helper inside to spot for leaks. The area around the skylight should also be wetted down so that any leaks that may develop while you are working can be found and repaired.

> A word of warning: arrange the water test so that you will not have to walk over any area which has been wet down with the hose. The water makes the roof very slippery and hard to walk on.

If you do find a leak in the flashing area, you should first check to see if any of the roofing nails that were used to secure the roof covering have punctured the new flashing. The way to fix this kind of leak is to remove the nail and cover the hole with roof cement.

If you have a leak in the flashing area but find that no nails seem to be puncturing the flashing, then the problem is insufficient coating of roof cement on the roof surface. Recoat the whole area where the flashing meets the roof material with additional cement.

The only other area where a leak may be found is where the Plexiglas has been joined to the curb. Check under the aluminum angle and make sure there is enough silicone caulk between the skylight material and the aluminum flashing. Pay special attention to the corners where the aluminum angles meet.

Framing the Shaftway

After the hole in your roof has been filled in with a skylight, it is time to complete the skylight shaft. Of course, if your skylight was built in a roof which also functions as the ceiling, you will not have to be concerned about framing in anywhere; there will be no shaftway. But if your skylight is set in a roof which is *not* the ceiling, then your construction requires a shaft between the ceiling and roof. You will have to cut a hole in the ceiling; this opening will have to be framed in like the one that was cut in the roof. Once you have cut the hole in the ceiling, you are ready to begin building the small walls that will form a shaft from the roof skylight to the ceiling opening.

The material for framing the shaft walls should be 2" x 3" studs. The frame will help stabilize the construction of the

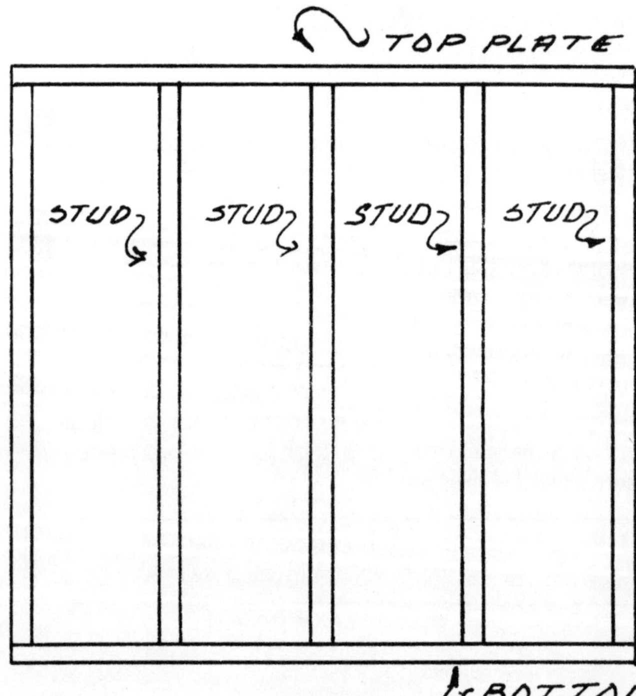

PARTS OF A TYPICAL STUD WALL

TWO 2×3'S NAILED TOGETHER WITH A SPACER OF ½" PLYWOOD SANDWICHED BETWEEN MAKES A GOOD CORNER BUCK
PLAN OF 2×3 STUD WALL SHOWING DETAIL OF CORNER BUCK. CORNER BUCK SERVES AS A NAILER FOR THE WALL COVERING

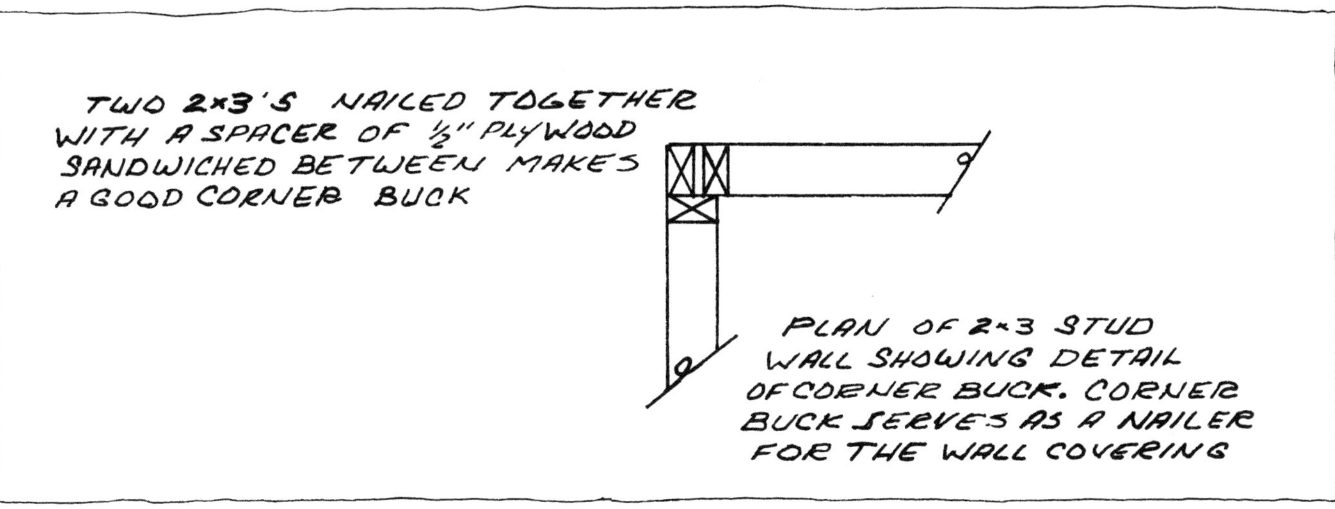

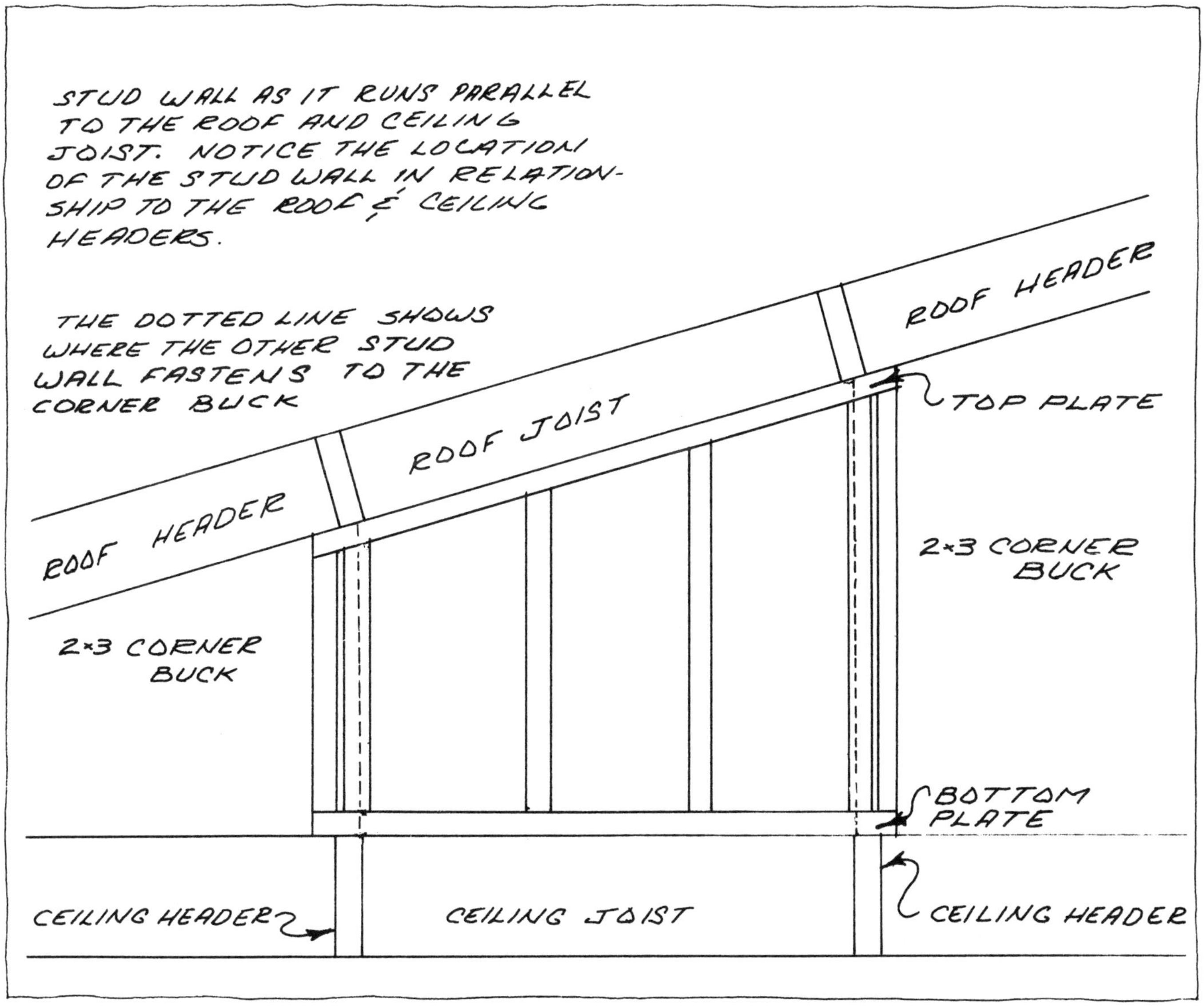

STUD WALL AS IT RUNS PARALLEL TO THE ROOF AND CEILING JOIST. NOTICE THE LOCATION OF THE STUD WALL IN RELATIONSHIP TO THE ROOF & CEILING HEADERS.
THE DOTTED LINE SHOWS WHERE THE OTHER STUD WALL FASTENS TO THE CORNER BUCK
ROOF HEADER
ROOF JOIST
ROOF HEADER
TOP PLATE
2×3 CORNER BUCK
2×3 CORNER BUCK
BOTTOM PLATE
CEILING HEADER
CEILING JOIST
CEILING HEADER

skylight and serve as nail blocks for the shaft wall covering and any electrical boxes.

The stud walls should be flush with the ceiling header and be set back from the roof header by the thickness of the wall covering. See Figure 12.

As you start to erect the shaft walls, you will notice that there are two types of walls needed. The first type of wall is one that runs perpendicular to the roof and ceiling joists.

> Note: If your roof has no pitch, then you will need only the first type of wall.

You can see by the drawings that there is a problem

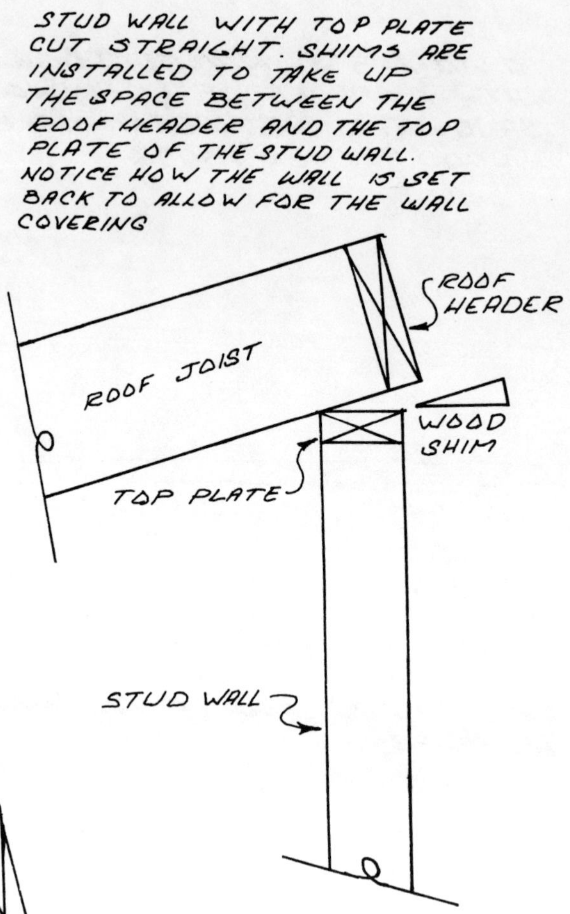

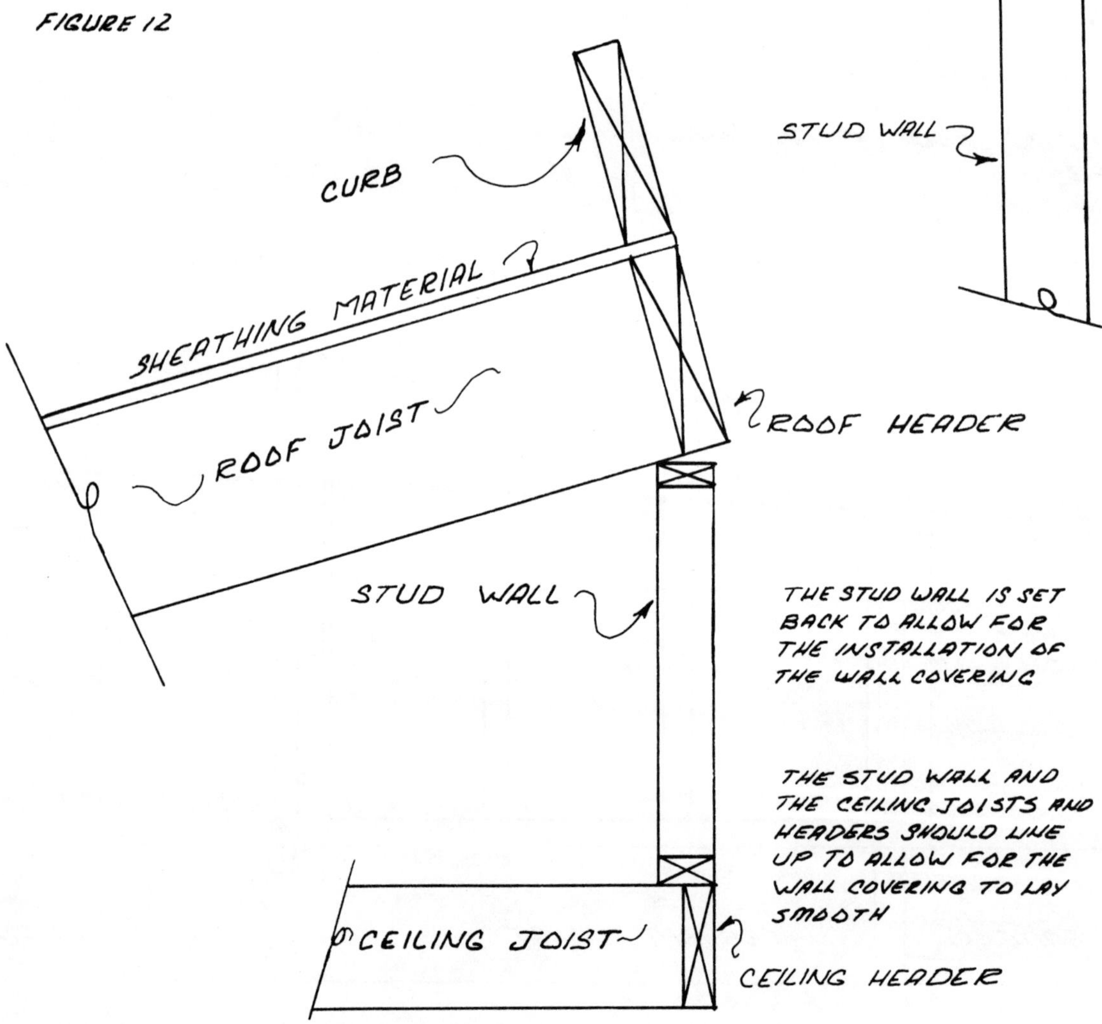

where the top plate meets the header. The gap between the stud wall and the roof rafter should be shimmed out so that the top plate is snug with all the roof joists.

The other type of wall will run the same way as the roof rafters and ceiling joists. The easiest way to build these walls is to assemble them on the floor and install them as separate units. For the angled walls, take the measurements of the length of the wall at the base and cut a 2" x 3" to this length. (Note that the actual measurements of the so-called 2 x 3 are 1½" x 2-5/8".) Measure the high and low points of the shaft wall and cut 2 x 3's to these measurements. Lay out the wall studs, including any extra studs which will be needed for nail blocks. Use the base plate as a guide for positioning the 2 x 3's.

Measure down 1½" on both the high-point stud and the low-point stud. This one-and-a-half inch space is to allow for the top plate. Using your chalk box, strike a line between these two new points. This line represents the angle and place at which all of the studs are cut.

After the studs are cut to the proper angle and size, nail the wall pieces together and install the wall in its proper position.

The only part of this construction that may present some problem is the corner. You have to be careful when you erect the wall that there is some surface to fasten the wall covering to. Check the drawing to see some of the ways to build these "corner bucks."

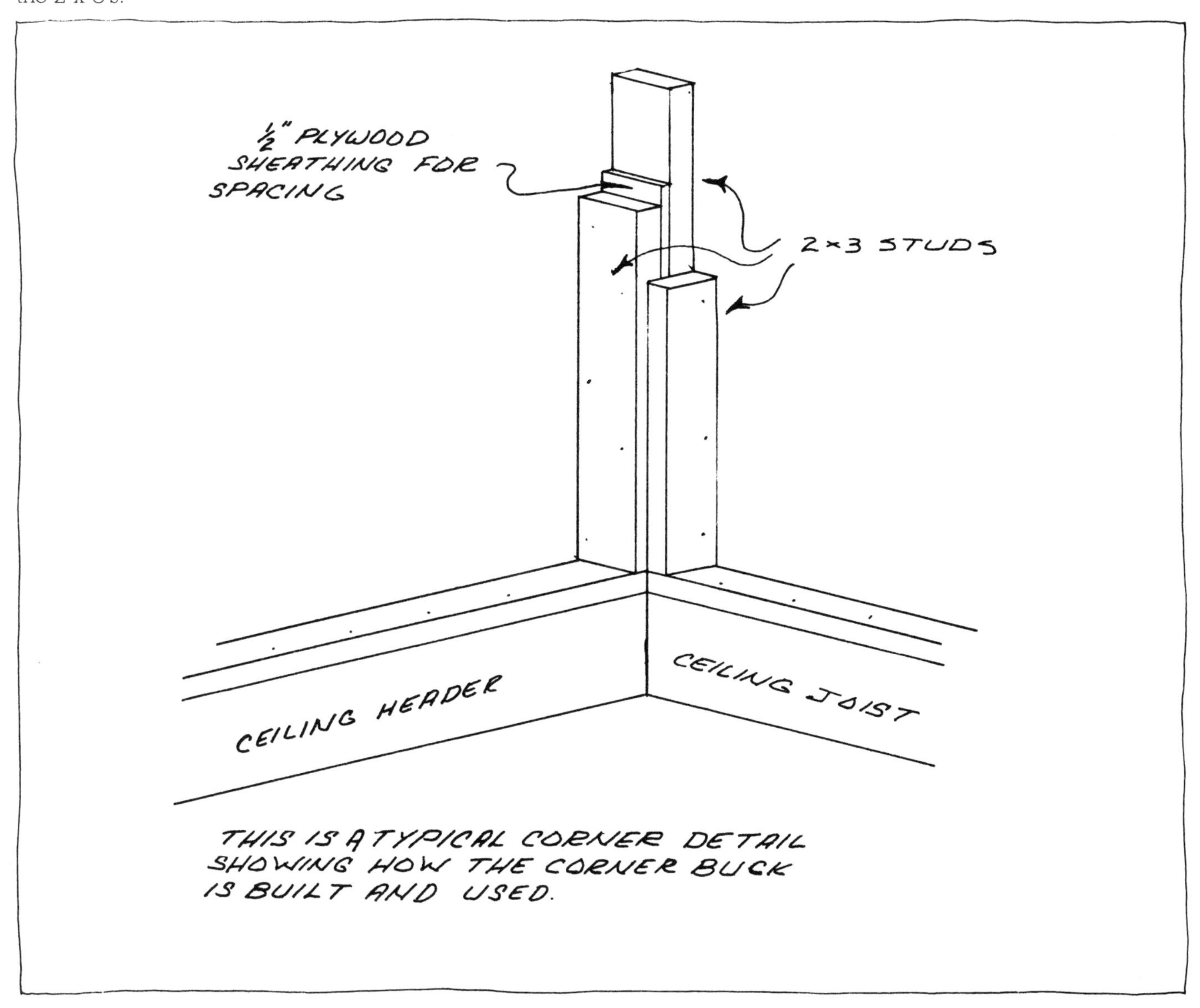

Electric Work: Installing Power for Lighting

If you need or want any electrical outlets for lighting in your skylight, the time to put them in is after the framing has gone up but before the wall covering has gone on.

There are two ways to bring the power up to the skylight. The first way is to build what amounts to a permanent extension cord; the other way is to splice into a junction box. Junction boxes are located throughout the attic.

The things you will need to supply your skylight with electricity are:

Cable — Known in the trade as Romex. It is a plastic sheathed cable with two conductors and a ground wire. The conductor wire comes in different thicknesses, or gauges. For your purposes, a 12-gauge conductor is the right wire. When you buy cable you should ask for "12-2 Romex," which means Romex cable with two 12-gauge conductor wires.

Electrical boxes — These fabricated steel boxes, which come in several shapes (square, octagonal, or rectangular), are supplied with stamped knock-out plugs so that the cable may be fed into the box easily. The boxes are of several sorts. Some of them have removable sides so that they may be joined (ganged) together. Others have specially designed brackets to allow them to be nailed or screwed to a stud or joist. Some have drilled holes instead of brackets. These holes are used to fasten the box to the stud or joist with 1" #10 sheet metal screws.

Cable clamps — Cable clamps are used where the Romex enters the box to provide a smooth surface (rather than the sharp edge of the box) for securing the cable.

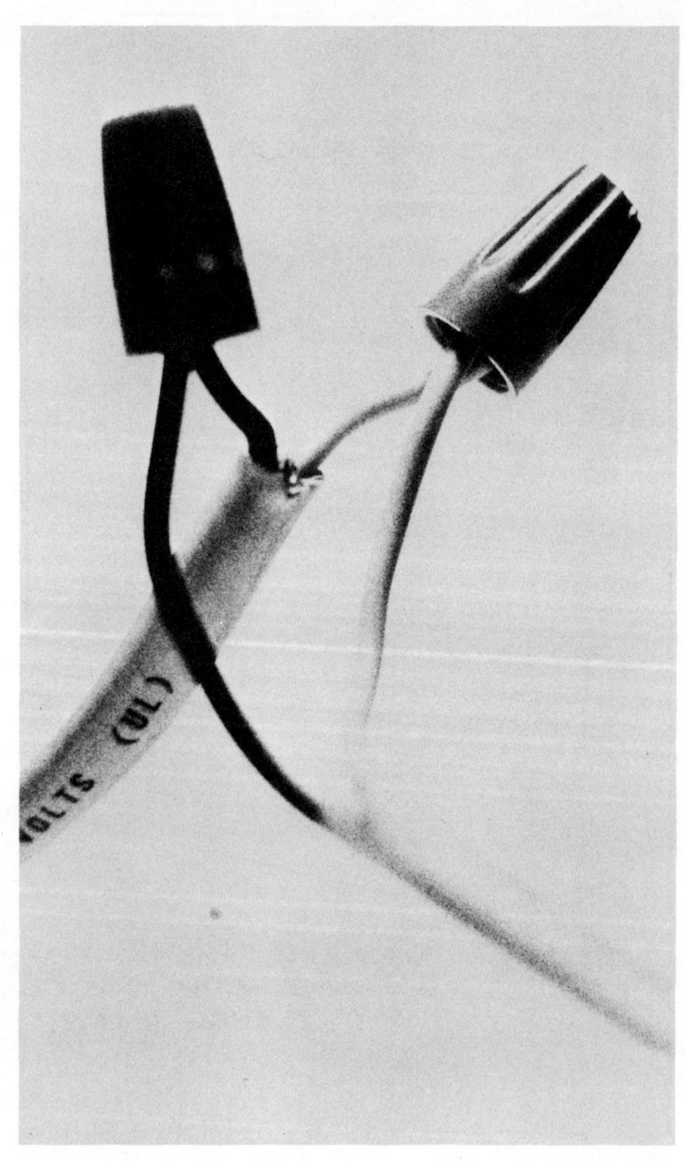

WIRES SPLICED TOGETHER WITH WIRE NUTS.

SOME OF THE MATERIALS NEEDED FOR ELECTRIC WORK.

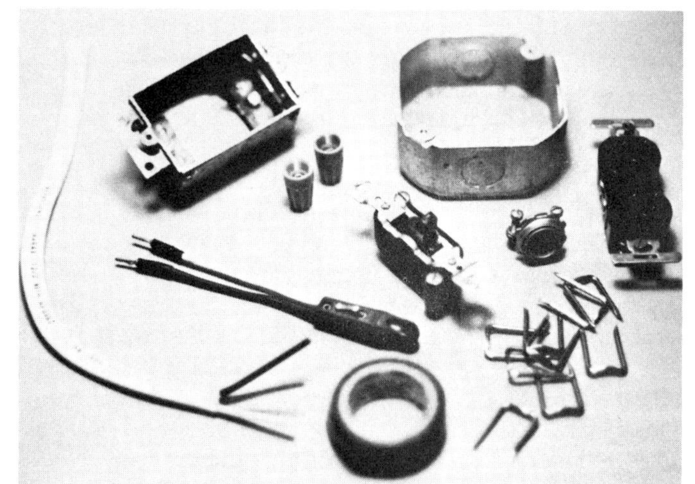

THE COMPLETED RECEPTACLE BOX. NOTICE THE USE OF THE CABLE CLAMP.

THE RECEPTACLE BEFORE IT IS ATTACHED TO THE BOX.

Wire nuts — These are funnel-shaped, metal-spring encased in heavy plastic. They come in three standard sizes. You will need the large size, which is good for splicing three wires together.

Receptacles and switches — Receptacles come in either a single or duplex style. They are constructed to receive a plug from an appliance or light fixture. Switches are devices used to connect or disconnect electrical current from its power source.

Plugs (in case you don't want to tap in at a junction box) — These are used to connect the appliance (in this case the skylight outlets) to the power source.

Now you are ready to figure out just what the electrical requirements of your skylight are. In the skylight that I built in my bathroom, I wanted a light that would not blind me when I had to turn it on after just having woken up. The way I solved this problem was really easy — and the best part is that it was incredibly cheap (about $2.50). I bought two *reflector hoods* with simple push-button light sockets. These reflectors come in a variety of shapes and sizes. The ones I bought were about 4" in diameter and about 7" deep. I cut a hole in the wall covering of the shaftway to measure exactly the outside diameter of the reflector hood and pushed the hoods into the holes. The hood has a lip at the edge that works as a stop and also makes the job look very finished. Since these lights come standard with plugs, I just installed a single rectangular box with a duplex receptacle for my power source. Here I did not have to mount the box in the wall, since the

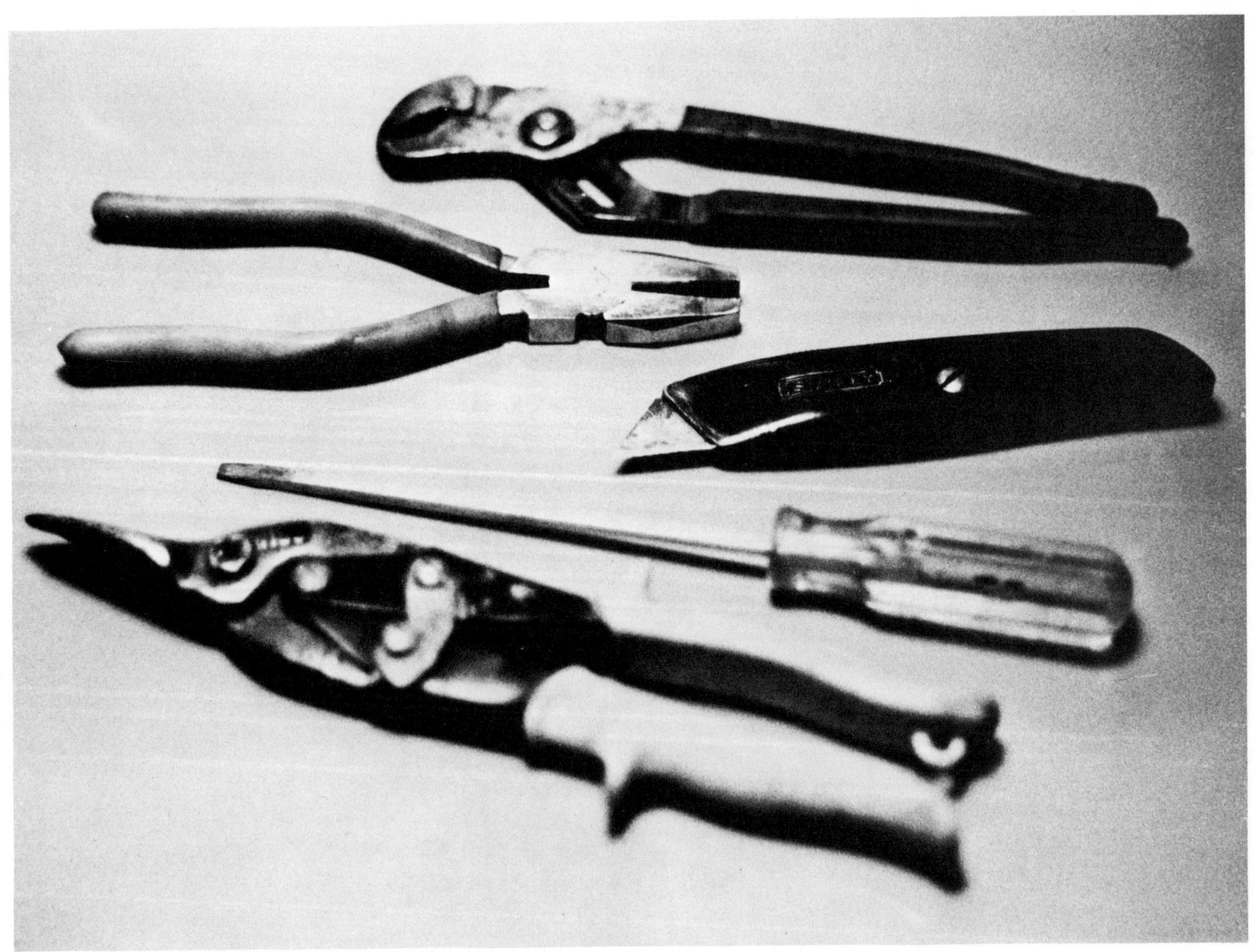

SOME OF THE TOOLS FOR ELECTRIC WORK.

plugs for the lights were inside the attic area. Make sure your power source is located such that you can reach inside the shaft wall to plug these lights in. The drawings in Figure 13 will show exactly how the installation was done.

Using the shaftway to conceal your light source, you can come up with some very appealing lighting designs. For example, if you have installed your skylight in a hallway, by adding some *flush mounted* lights you can effectively create the same atmosphere at night as you have in daylight.

After you have decided what you will need in the way of electricity, you must decide where you will tap into the house power. Remember, when you put a light in your skylight you must have a switch to control the circuit.

The basic plan for both the permanent extension cord and the junction-box splice is the same. The circuit goes from an outlet box with a duplex receptacle — fastened to a convenient location near the skylight — through the cable to a switch (to control the lights) and from the

FIGURE 13

THE REFLECTOR HOOD IS PRESSED INTO A HOLE CUT IN THE SHAFT WALL COVERING

REFLECTOR HOOD

DUPLEX RECEPTACLE

switch to the power supply. Look at the drawing carefully to see the sequence in which the circuit is built. The most difficult part of the whole thing is finding a location for the switch. There is no problem running the cable through the attic and the shaft walls, but things get a little tricky when it comes down to the existing wall where you are going to mount your switch box.

The easy way to do this job is to locate your switch box in a stud wall. This way you have a ready-made channel for the cable to run through. First, cut a hole in the wall where the switch box will be located. Then, as the drawing shows, drill a hole in the top plate of the wall and feed your wire through this hole. Use another piece of wire to fish it through the hole you have cut in the wall for the switch box. Install the switch box in the wall,

FIGURE 14

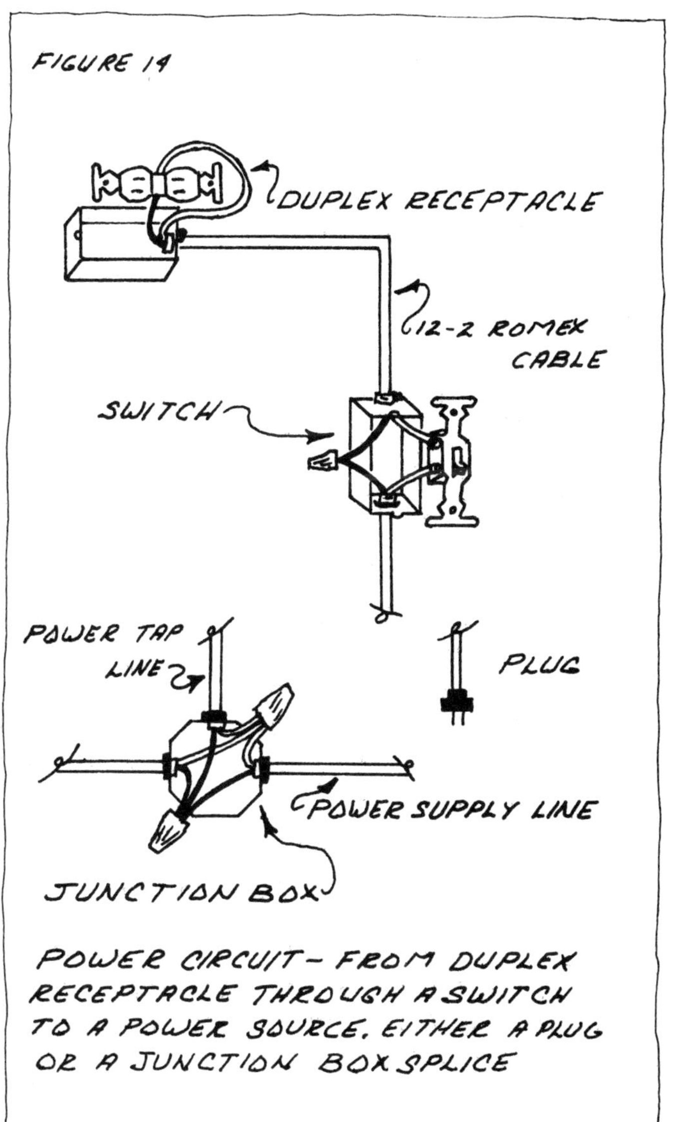

POWER CIRCUIT – FROM DUPLEX RECEPTACLE THROUGH A SWITCH TO A POWER SOURCE, EITHER A PLUG OR A JUNCTION BOX SPLICE

EXAMPLE OF AN INCANDESCENT LIGHT FIXTURE RECESSED INTO DRYWALL.

feeding the cable through and securing it with a cable clamp.

> Note: Use the right style box for the walls you have. There is a specially made box for drywall and one also for plaster lathe.

After you have installed the switch, following the wiring diagram in Figure 14, then you can attach your cable to the power supply. Here is where the difference between the two power-supply methods comes in. With the permanent extension cord type, the power tap comes from plugging the cable into an existing outlet. This is the simple way to do it; but my personal preference is for the junction-box splice method because it gives you a top quality job and is one of those small details that makes the whole job more professional. It may be a little difficult, but if you follow the steps listed below you should have no problems in splicing into a junction box.

1. Locate a junction box in your attic. Try to pick one that is in a convenient location.

2. Take off the cover and separate the groups of wires.

3. Here you will need a helper and a flashlight. *Have your helper shut off all the circuits or fuses in the house.* When all the power is off remove the covering of the spliced wires. Add another cable clamp by punching out one of the knock-outs. Slide your cable in. Strip off

TRACK LIGHTING MOUNTED ON THE SURFACE OF DRYWALL.

the protective covering and bare the ends of the wires.

4. Join each wire to its appropriate group—match the colors (black to black, white to white)—either with a larger wire nut or a good wrapping of plastic electrical tape.

5. Push the bunches of wire back into the junction box and put the cover plate back on.

6. Have your helper turn the power back on.

If you decide to use the plug method, then instead of running the cable to a junction box after it passes the switch, just run it to the nearest receptacle. Since you will have to mount the cable on the outside of the wall to reach the plug, you may as well run all of the switch wiring on the surface.

If you can't find a stud wall to run your cable through, and the only wall you have to use is a brick and plaster one, then you have to (1) cut a small channel in the plaster to allow the cable to run down, and (2) cut a hole in the plaster deep enough for a specially made box (one that is not nearly so deep as a standard outlet box). Repair the wall with patching plaster. Remember, you'll have to sand and repaint after this part of the job is done.

Although it may seem like a lot of work to install lighting in your skylight, the finished product is well worth the relatively little time spent doing it. Always remember to use your common sense when working with electricity.

Shaftwall Insulation

After you've finished the framing and the electric work, now is the proper time to insulate your skylight shaft. *Roll insulation,* as it is called, is available in two types. One of the major differences between them is the paper backing that holds the fiber glass insulation material together. The insulation comes with either a foil back or a kraft paper backing. The foil back is more expensive than the kraft paper backing, but insulated a little better in that it helps reflect the heat back into the house. But either kind will work quite well.

Note: If your attic is not already insulated, now is the time to do it. You will really be amazed at the difference. And remember, insulate in every nook and cranny of the room while you're at it.

Another difference is in thickness. Roll insulation comes in two standard thicknesses—3½" and 6". If you have a choice, *the thicker the better.* It also comes in two different widths, 15½" and 24". Your choice here depends on the width between your studs. Once you know the width you need, the best way to cut the stuff to the necessary length (the height of the shaftwall) is with a mat knife guided by a straightedge. Insert the strip of insulation between the studs. The insulation comes with a flap along both its edges; just staple the flaps to the studs and the job is done.

Covering the Shaftway with Drywall

Now that you have the skylight installed, the walls framed, insulated, and supplied with power, you have to decide what you are going to cover the walls with. In most houses the ceiling is likely to be either plaster or drywall. In either case, putting drywall (sheetrock, plasterboard) on the new framing can be an economical and attractive solution to the problem of finishing your walls. Some other alternatives for finishing the framed shaftway are suggested in the next section.

Before the work begins let's go over the tools that will make the drywall job easier.

Drywall trowels — I use three trowels. (1) The first trowel is a five-inch trowel used for scooping drywall paste into the pan and for doing some first-coat work. (2) The second trowel is my general all-purpose model. It is twelve inches long and has a very flexible blade. This trowel is used for taping seams and for applying second and third coats of the paste over the nails. (3) My third trowel is for corner work. It is bent at a 90° angle and has a flexible blade. It makes inside corner work go like a breeze.

Mud pan — This pan is used for carrying small amounts of drywall paste so you won't have to dip into the large can any more than necessary.

Straightedge — There is a square available with a four-foot leg; but unless you are going to do a lot of drywall work, the money can much better be spent elsewhere. The solution is a nice straight piece of 1" x 4" pine. For long lines, a chalk line is the tool to use.

Drywall saw — This tool is a must ($1.49) if you have to cut openings in the drywall for electric boxes. It makes the job quick and the results satisfactory.

Sandpaper — You will need some sandpaper to smooth the hardened paste over the drywall. You should use an 80-grit paper.

TOOLS NEEDED FOR DRYWALL WORK.

A PROPERLY DIMPLED DRYWALL SEAM.

APPLYING THE FIRST COAT OF DRYWALL PASTE IN PREPARATION FOR DRYWALL TAPE.

Once all these tools are assembled you can get to work. The first thing to know is how to cut the drywall. Use your square or a straightedge as a guide for cutting through the top layer of the drywall.

> Note: You will notice that the drywall is composed of three layers of material. The top and bottom are a cardboard-like product which gives substance to the gypsum that makes up the large center layer.

First use a mat knife to score the top layer and break the center gypsum layer; then snap the drywall back to get a nice even break. Keeping the drywall in this position, score its back side. Pull the drywall sharply the other way and the piece will break cleanly.

After you have measured and cut the drywall and are ready to nail it up, be sure you have all the nailers in place. (Check corners carefully.) Special drywall nails are the proper fasteners to use. They are blue in color (a coating to inhibit rust), have a large head, and are ridged to keep from pulling out. When you are hanging the drywall the nail should be driven so that the hammer has "dimpled" the drywall (i.e., the head of the nail is indented slightly beneath the surface of the drywall). When you dimple the nails, try not to break the top paper of the drywall. Dimpling in this way will enable you to cover the head of the nail with drywall paste so that your finished surface will be completely flush. The tapered edges of the drywall should be butted together; this will make the job go a lot easier.

The only thing left is to cover up the seams (which is not as difficult as it may seem). Drywall paste comes two ways: ready-mix and in bags. I think that the best thing to buy is the ready-mix variety—there is absolutely no hassle and much less mess and clean-up.

> Note: One of the easiest ways to guarantee a good drywall and paste job is to keep your trowel free from dirt and particles of drywall. The advantage of using a pan is that if the paste does become dirty only a small amount needs to be thrown away, and it *should* be thrown away.

Drywall tape is used with the paste to cover up the seams of the drywall. Using the 4" trowel, the first step is to apply a coating of paste to seams. After the paste is down, center one end of the tape over the joint and press the rest of the tape firmly into the paste. As you pull the trowel along, the paste should ooze out slightly from under the tape. Don't press *too* hard or all the paste will come out and you won't get a good bond between tape and drywall. And don't put on so much paste that you end up covering the tape on the first coat. By the time all three coats are on, the tape will have neatly disappeared.

After the first coat is dry (about 24 hours), a second or filler coat should be applied. Using the same material with a wider trowel, spread the paste a couple of inches on both sides of the taped seam, feathering out toward the edge. *I repeat again: don't use too much paste. It is*

APPLYING THE DRYWALL TAPE.

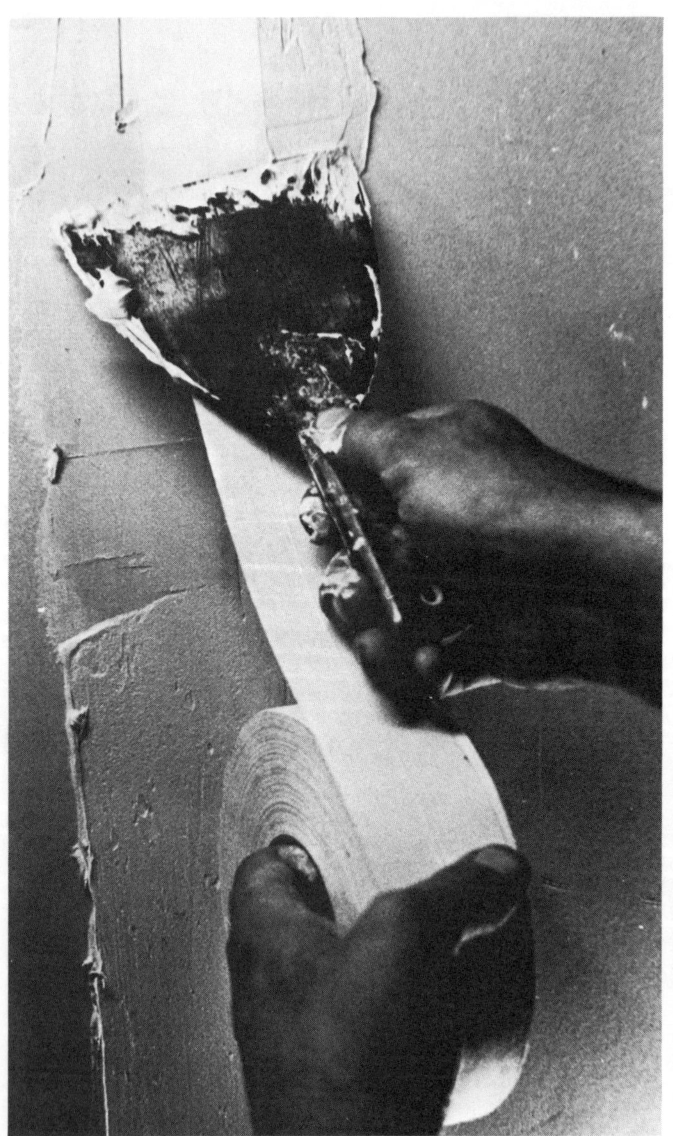

APPLYING THE SECOND COAT OF PASTE.

USING THE WIDE-BLADED TROWEL TO APPLY THE FINISH COAT.

APPLYING TAPE AND PASTE AT THE CORNER, USING THE TROWEL DESIGNED FOR THIS PURPOSE.

easier to fill in a dip than to sand down the lumps when the paste is dry. When the second coat is dry it should be sanded lightly, and then the third or trim coat can be put on. The third coat is used to fill in the recesses and to smooth the joint to a finished quality. The final sanding should take care of the minor imperfections left in the third coat.

The only thing you will have to watch out for is the seam that is not tapered. If you have cut seams, or if you have cut the short edge of a full sheet, there will be no tapered end. Here the trick is to make the feathered edge wider so that the hump created by lack of a tapered edge will disappear. To do this, work through the same steps as for the tapered edges.

To tape an *inside corner,* measure the correct amount of drywall tape and fold the tape in half. Apply a first coat of paste on both sides of the corner. Position the tape in the corner and use the corner trowel to press down the tape securely. Use the wide trowels to feather the seam, as in the tapered edge.

The *outside corner* is made with a product called metal-cornerbead. This bead is nailed in place with drywall nails. The cornerbead is used so the corner won't be damaged when it is bumped. Feather both sides of the corner with increasingly wider coats of drywall paste; sand smooth in between coats.

Nail dimples are finished over by the same three-coat process, with sanding in between coats.

BITS OF GLASS, CERAMIC, MIRRORS, AND METAL ARE USED IN THIS WALL SCULPTURE.

Other Kinds of Wall Coverings

There are other alternatives besides drywall for finishing your skylight shaft. One of the cheaper wall coverings that could be put up is cleaned packing-crate lumber. Think twice before you scoff at this idea. The walls of your skylight shaft are not very long or wide. Packing-crate lumber would be convenient for finishing because it is generally available in appropriate lengths.

If packing-crate lumber is not your style, then you might consider using random-width pine boards or uniform size pine boards (one-by-sixes, for example). Pine boards might be layed on the diagonal for an attractive design. You can buy either #2 *white pine,* which has some knots, or *clear white pine,* which should have no knots. It all depends on which appeals to you more. Either would be fine as a wall covering.

If you'd prefer not to lay down all those planks, you could use paneling. Paneling comes in many different grades, colors, and textures. The cheapest paneling has a photo overlay on a Masonite or mahogany panel. I recommend that you avoid this kind of paneling because it has a tendency to look like what it is—sloppy and cheap. Veneer paneling has an actual wood covering over the mahogany panel. This kind of paneling has the look and feel of real wood and, if the color is right for you, is a very suitable covering for your shaft walls. One make of paneling which I would like to mention specifically is Georgia Pacific Texture 1-11. This has an exterior plywood which looks like planking and is a real great way to make the skylight shaft look beautiful. Whatever your paneling interests are, check it out at the lumber yard. They usually have plenty of samples to show.

A final word about choice of wall covering. Remember that anything that is used as a wall covering can be used as a covering for the shaftway—and some things may work for the shaft that could not work over a whole wall. For an unusual shaft, you might like to use carpeting; or fill the skylight with plants and wallpaper the shaft in a jungle scene. The shaftway may be the place where you could use that unusual color paint that you've been storing in the utility closet for just such an occasion.

After you have decided which covering to use and are ready to install it in the skylight shaftway, here are some things you should know before the actual construction begins.

The wall covering should run to the underside of the roof header, as shown in Figure 16. Unless you are

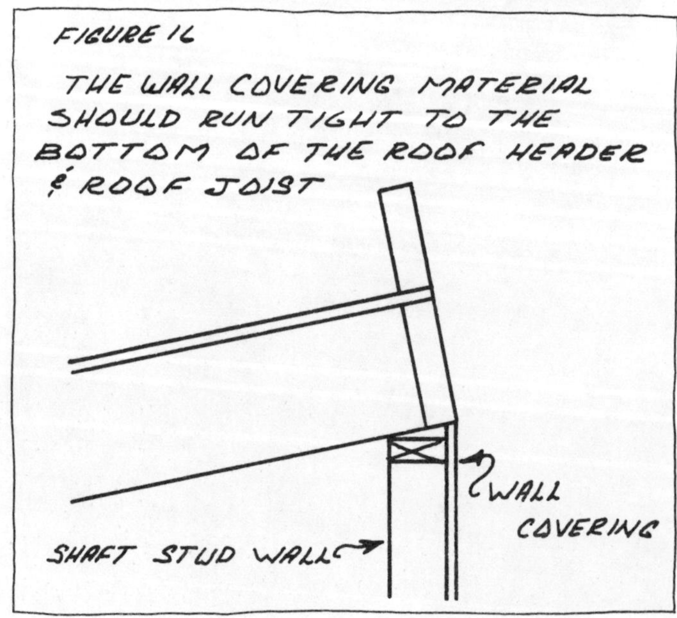

SOME WALL COVERING SUGGESTIONS.

CORK

FABRIC

using drywall and metal corner bead, the corner where the ceiling meets the shaft wall requires some special treatment. Outside corner molding (purchased from any lumberyard) is generally used on those walls which have been paneled. See Figure 17.

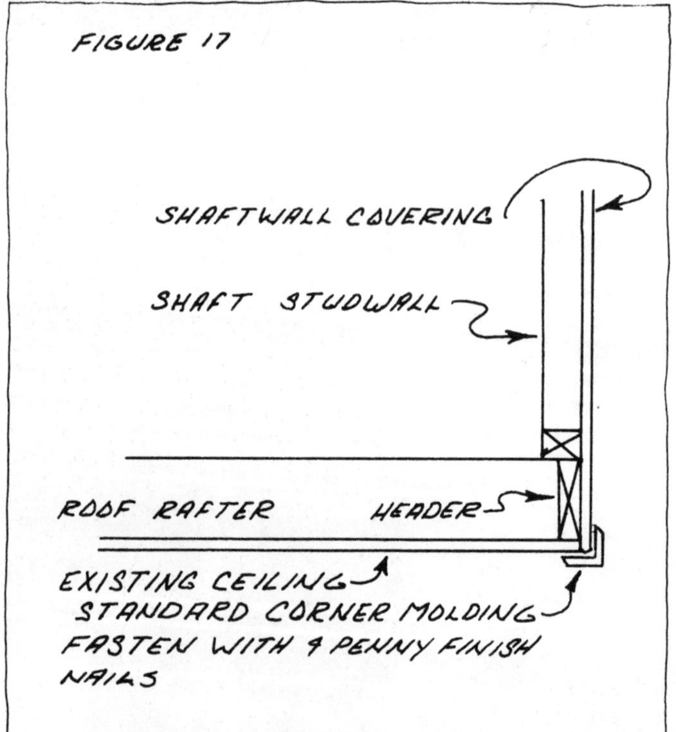

If you have used planks of some sort, then the skylight could be trimmed out with the same material (see Fig. 18). The four inside corners can be dealt with in a couple of ways. The first is to cut the wall covering so tight that you would get a good seam at those corners. (A good seam is one which has no gaps between the two pieces, and the line created by joining the two should be straight.) Alternatively, for paneling you can buy pre-milled molding made specifically for the job. If you are using planks, then you can finish the corners with narrow strips of the same material that was used for the wall covering. See the drawings in Figure 19.

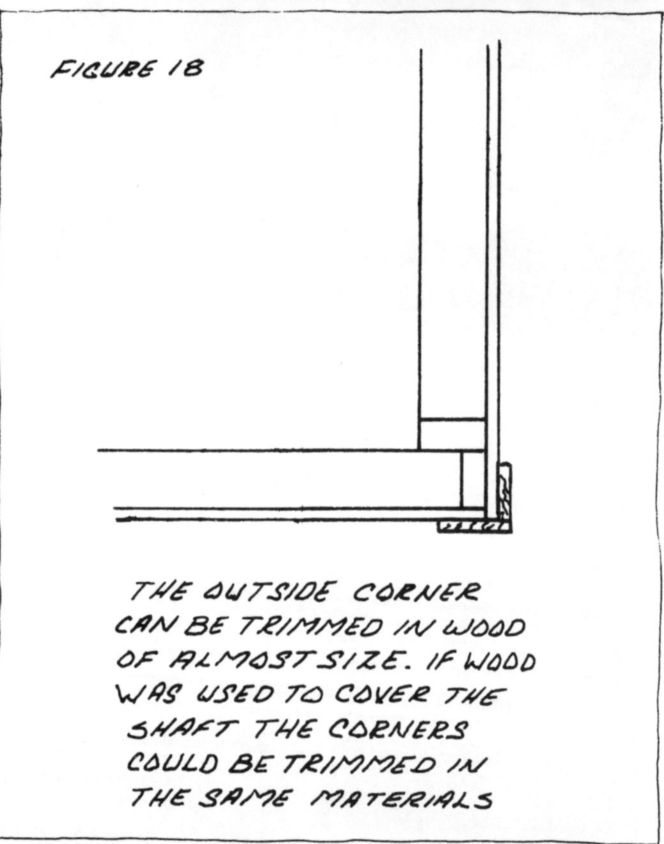

WOOD

SHAFTWAY COVERED WITH WOOD AND PAINTED.

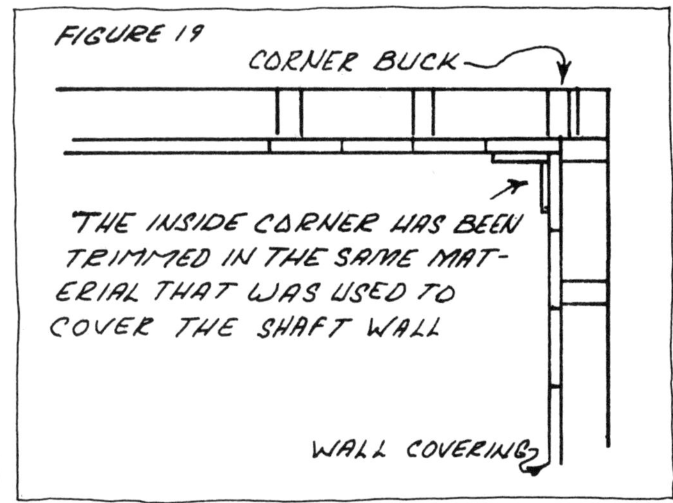

Internal Storm Windows

In my opinion, thermal glass *may* be the most efficient insulation against heat loss. Thermal glass is two pieces of glass joined together at the edges with a rubber sealant. The vacuum in the space between the two pieces is what insulates the window. *However,* thermal-pane glass is very expensive, especially in the larger sizes. So the thing to do is make something which acts like the thermal glass but does not cost the money. The solution is a simple one. Make what amounts to an *internal storm window,* which in effect works on the same principle as the thermal glass. The space between the two frames of glass acts as an insulator between the outer temperature and the inner temperature. The reason for the vacuum in thermal glass is that the fewer air molecules to transmit outside to inside, the more efficient the insulation will be.

That is the why of insulating the skylight—and here is the how.

There are two easy ways to build this internal storm window. The first thing to do in both cases is to nail a 1"x2" pine strip on edge around the inside of the curb so that the top of the strip touches the skylight. Figure 20 shows exactly where this pine strip is to be nailed.

FIGURE 20

SKYLIGHT MATERIAL

INTERNAL STORM WINDOW

1×2 PINE STRIP NAILED AROUND PERIMETER WITH EIGHT PENNY COMMON NAILS

Note: If you know for sure you are going to use a storm window, the pine strip should be nailed on when the curb is installed.

After the pine strip is nailed on, then just staple some plastic sheeting to it; if you want, you can put a cover strip over the staples, as in Figure 21. Remember that, although this does let the light in, the plastic sheet is not completely transparent. This may be all right in a circumstance where the skylight is used primarily for light, like in a workshop or studio.

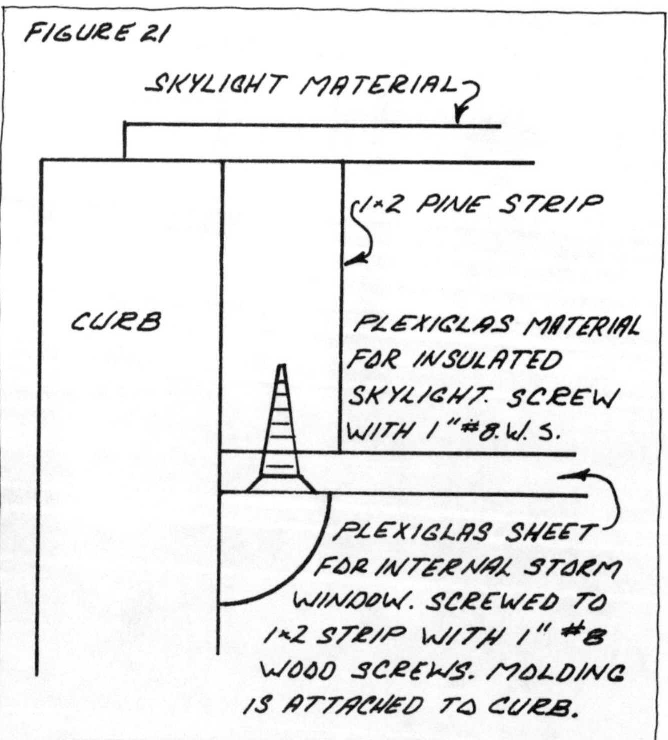

There is another method that will solve the insulation problem and still let you look out at the world with a clear view. After the pine strip has been nailed in place, cut a piece of Plexiglas to the same measurements as the inside of the curb. Then drill holes in the Plexiglas about every 12 to 14 inches around the perimeter. These are for wood screws that will fasten the Plexiglas to the bottom of the pine strip. An extra thing to do before you screw the storm window in place is to install a piece of weather stripping around the bottom of the batten. For weather stripping, you can use felt strips, which are nailed in place with a small brad, or you can use strip rubber, which comes in a roll and has an adhesive back. Peel the protective tape off the strip and press it into place. With both types of weather stripping you can drill the pilot holes for the storm window right through the stripping.

You should use 3/16" Plexiglas for skylights larger than nine square feet. For the smaller skylight, 1/8" will work fine. Remember, it is not the thickness of the material that insulates the skylight, but the layer of trapped air.

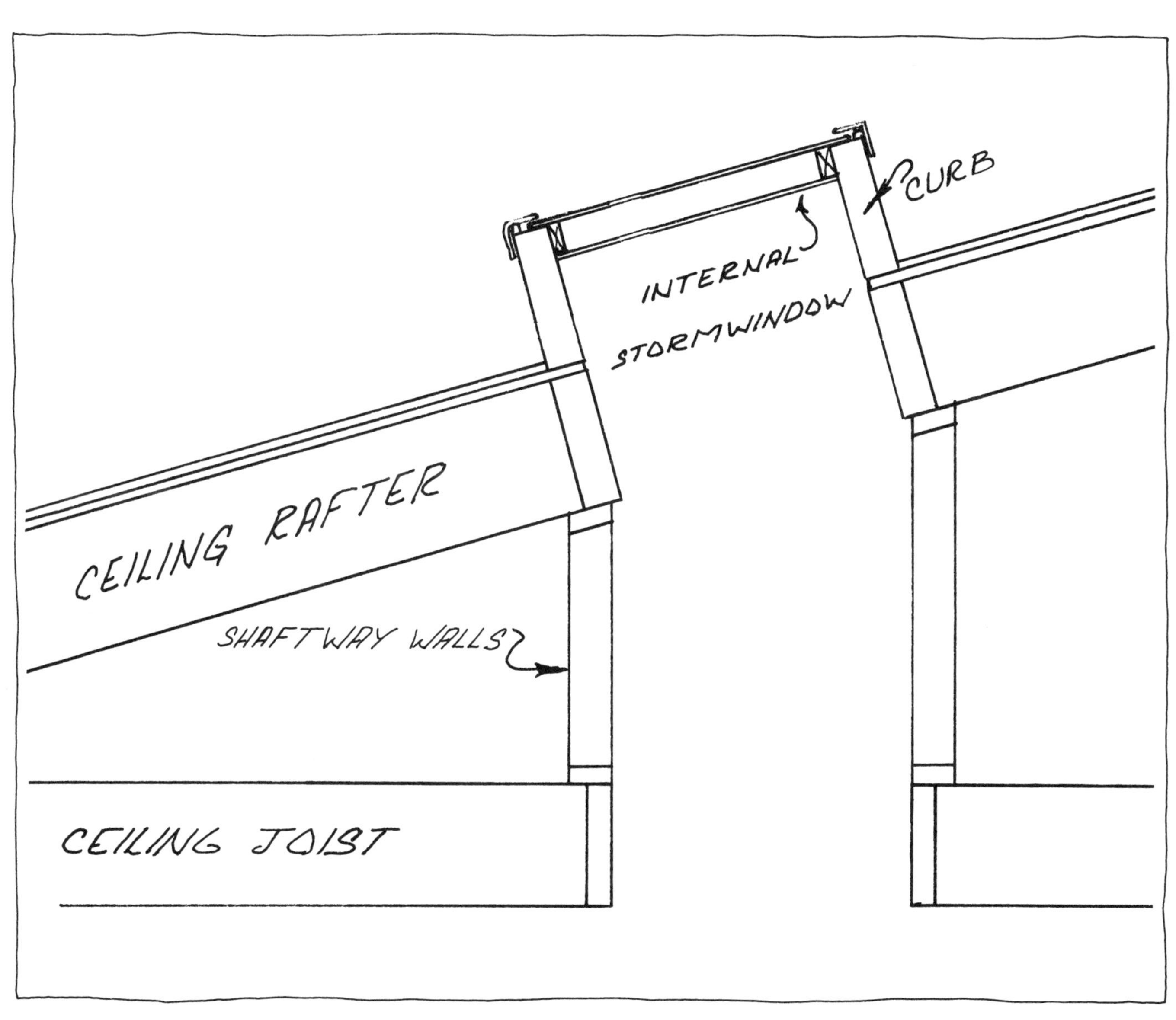

Plants Under Your Skylight

Light is the key factor in the growth of house plants. It is the essential ingredient in the process of photosynthesis, by which a plant produces its own food. Different species of plants require differing light levels, but they have in common their reaction to insufficient light: a cessation of growth. Unless a plant is supplied with the approximate amount of light it needs, it will soon die.

Yet most of us live in rooms in which the supply of sunlight is severely limited; perhaps nearby shrubs, trees, or buildings keep much sunlight from entering. Perhaps you just don't have windows with a southern or eastern exposure: northern and western windows receive very little direct sunlight. Whatever the cause, the result is the same. While you can grow a variety of plants capable of doing well in relatively low-light situations, you'll be unable to raise many (including most flowering plants) that do need plenty of direct sunlight. You could accept the situation, and make do with the plants that will survive in your rather dim rooms. Or you could install an artificial light unit, consisting of a series of fluorescent tubes, under which many kinds of sun-loving plants can be grown. Or you might consider installing a skylight.

There are many good reasons for adding a skylight to a room. Admittedly, doing it to increase your selection of house plants may be far down on the list. But it *is* a valid idea. If you own a home, if you like plants, and if you can't grow all of the species you'd like to because of a lack of light, you should certainly consider adding a skylight. And if you've decided to add a skylight for other reasons, you shouldn't neglect the chance of adding plants as an afterthought: plants and skylights can be combined to produce stunning decorative effects.

You can suspend plants in baskets around the edges of the skylight. You can group large plants beneath the light. You might even want to build a water garden. Given the brilliant light the unit should make available to the room, you should be able to raise even the most light-hungry species of plants. Don't limit the addition of plants to the area beneath and around a skylight. When you knock a hole in the ceiling and let the light in, the entire room will be measurably brighter, even though much of it will not receive the direct rays of the sun. This general improvement in the level of light in a room makes it possible for you to add a variety of plants that don't need direct sunlight, but do need bright indirect light. No longer will you need to be limited to those few hardy species that can prosper in dim locations.

After you've installed the skylight, watch to mark out the path of sunlight across the room, from morning until sunset. Across that path, you can arrange the plants that need bright sunlight. Around it you can place those plants needing bright but indirect illumination. And in the least bright spots, you can keep those hardier plants that do well with only moderately strong light.

Because light will be moving across the room, in accordance with the path of the sun, you should not have to worry that plant leaves or blossoms will be scorched by the rays of the sun: direct sunlight won't linger in any spot long enough to cause such damage. However, the area directly beneath the skylight may become much warmer than other areas of the room, and excessively high temperatures can harm plants.

Take a temperature reading directly below the skylight, when the sun is falling directly on that spot. Take another reading when the sun has moved on, illuminating some other part of the room, leaving the spot beneath the skylight lit by indirect light. Place the thermometer at the same level as the tops of the majority of plants involved. If the temperature does not decrease very much after the sun has moved on, you'll have to take some steps to protect the plants. Misting the plants frequently will help. What seems a problem in the spring or summer can become a benefit in the winter, as the extra amounts of warmth retained by the unit will be appreciated by any plants you're attempting to bring to blossom.

Hanging plants and skylights seem made for each other. Suspending plants in containers has only recently reemerged in a big way, but the idea now seems firmly settled in the marketplace. Containers are available in an astonishing array of materials, shapes, and colors. There are containers of unglazed or glazed clay, pots of brass or copper, plastic pots, and handwoven pots. There are also wire baskets, lined with strips of moss and then filled with soil. Such baskets drip a good deal of water after each watering, making them impractical for use indoors—unless you want to suspend them over moisture-loving plants, which will appreciate the frequent showers. When you go shopping for containers, buy only those that have a saucer attached beneath the drainage holes, to catch the run-off of liquid after each watering.

Containers can be suspended with any sort of strong material: nylon or hemp ropes, thick woven ropes, leather thongs, chains, or a strong length of wire. Brackets, ceiling hooks, or side hooks can be used to hang the container in place. In the case of skylights, ceiling hooks would be the best choice. Place the hooks along the edges of the skylight, so long as the material of the ceiling is strong enough to support the weight. If you judge it not to be, you'll have to suspend the containers from those beams nearest to the skylight. If you intend to suspend very large or very heavy containers, you'd best substitute molly bolts or toggles for the ceiling hooks. Another place to attach hooks for hanging pots is the studs of the shaft wall. Before any hooks are set in place, know the size of your containers and the width of the plants. Don't hang plants so closely together that their foliage overlaps, for such a practice will cause leaves to yellow or drop off, and may serve as a bridge for pests passing from plant to plant.

Of course, you can also arrange plants beneath the skylight. Without much difficulty, you can create a panorama of green life around or just below your skylight that will rival the most exotic jungle scene. One gardener I know has created a mini-desert, filled with many forms of cacti, just below his skylight. Another has installed a small pool with a variety of water plants growing in and around it. To stand by it, when the sun is falling through the skylight, and hanging plants are dangling from above, is to feel yourself in some tiny glade in a rain forest.

You needn't go to such elaborate efforts or expense. The point of such experiments is that almost anything is possible. A skylight is, in my opinion, close to a miracle-worker when it comes to plants. For a gardener, it makes almost all things possible. If you're installing a skylight, don't forget the plants. And if you're raising plants, please consider seriously the installation of a skylight.

Perhaps you'd like to know which plants in particular would be especially pleased to take a seat under your new skylight—and which ones would rather not. So here is some specific information of the light requirements of some common house plants.

Plants needing direct sunlight: Blood-leaf; Bougainvillea; Calla Lily; Chinese Hibiscus; Crown-of-Thorns; Donkey's Tail; Geranium; Oxalis; Shrimp plant; Swedish Myrtle; any species of Cactus.

Plants needing only some direct, but plenty of bright indirect, light: Amaryllis; Asparagus fern; Avocado; Jade plant; Pittosporum; Passion Vine; Rubber plant; Ficus; Wax plant.

Plants requiring little direct light: Aglaonema; many species of Bromeliads; Cast-Iron plant; Caladium; Coleus; Dracaena; Dumb Cane; English Ivy; Grape Ivy; Peperomia; Philodendron; Prayer plant; Spider plant; Umbrella Tree; Wandering Jew.

Many species of ferns, requiring protection from direct exposure to sunlight, do very well in hanging baskets.

TIRES TIRES

Places to Buy Things

One of the advantages of building your own skylight is that the materials are so easily available, and you need go only to three or four different places: a lumberyard, a hardware or houseware store, and a glass or plastic supply store will have everything needed to make the skylights described in this book. And you can locate everything at first by using the Yellow Pages—one of the most useful of all tools. Call before you go shopping; you'll be surprised at how much time and effort will be saved.

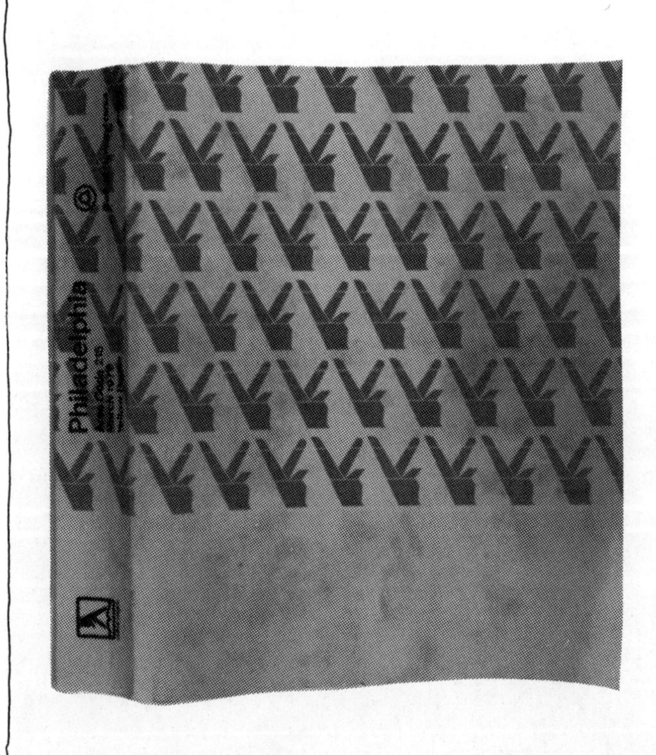

Skylight Manufacturers

I don't want to give a list of manufacturers of skylights. I don't recommend them; they cost three or four times as much as the ones you can build. *But,* if you want, you can obtain a list of manufacturers by going to the library (Business, Science, or Industry department) and asking for *Sweet's Architectural Catalog File for Design and Construction of General Building* (about 20 volumes, all indexed).

This set of books contains almost every maker of just about anything that is needed in the building trades. Before you do any purchasing, I would like to point out some features that the skylight unit you may be considering should have.

1. The skylight should have an inner and outer shell which would create the dead air space that is necessary to insulate the skylight.

2. The aluminum angle should have welded corners and should be of such thickness that it is rigid and firm.

3. The sealant for these skylights should be a silicone or polysulfide type. Ask to make sure.

4. Ask how the skylight is to be fastened to the curb. If the procedure seems too difficult, then look elsewhere, as there are superb skylights which attach to the curb easily.

Manufactured skylights come in a lot of different shapes. Some of the shapes available (other than flat) are the dome, box, square pyramid, and triangular pyramid. They also come in two basic colors—neutral gray and bronze—and in five densities for each color from light and dark. Some of these skylights can be purchased with a frame that allows it to be opened up.

The manufactured skylight comes in standard sizes to minimize the carpentry needed to install these skylights.

NEW MATERIALS ON OLD STRUCTURES SOMETIMES CREATE BEAUTIFUL EFFECTS, AS SHOWN BY THIS CONTEMPORARY SKYLIGHT BUILT INTO THIS 18TH-CENTURY HOME. ONE ADVANTAGE OF ACRYLIC DOMES IS THEIR CLEAN FINISHED EXTERIOR APPEARANCE.

THESE PHOTOGRAPHS SHOW THE EXTERIOR OF THE HOUSE IN THE WOODS (PICTURED ON PAGES 42–45).

Glossary of Terms Used in this Book

aluminum angle. Aluminum material bent at a 90° angle. Comes in any length up to 12 ft. and in a variety of widths.

amp rating. A means of measuring the power of electrical tools and motors.

batten. Narrow strip of wood nailed around the top of the curb. It is used as a spacer to maintain the correct distance between the aluminum angle and the skylight material.

bottom plate. The base piece of studding to which the vertical support members of the wall are nailed.

corner bead. Formed sheet metal strip, used on corners when hanging drywall before the drywall paste is applied. Its purpose is to reinforce the corners.

curb. The wooden framework, usually made of 2 x 6's, that is used to frame the skylight material and give it a suitable base.

drywall. A kind of gypsum board—a wall covering used on interior walls. Drywall is available in large sheets; 4' x 8' is most common; also comes in 4' x 10' and 4' x 12'. (See **gypsum board**.)

epoxy. Two-part adhesive which, when mixed together, forms a strong bond between the materials being joined.

flashing. Material used in roof and wall construction to protect against water seepage. Material is usually aluminum and comes in varying widths from 6" to 18"; it can be bought in any length.

framing. A system of joining lumber together by nailing to form the skeletal structure of walls, floors, and roofs.

galvanize. A way of treating steel—in this case nails—against rust.

gasket. Any material which is used to create an airtight joint between two materials.

gypsum board. Material made from gypsum sandwiched between two layers of heavy paper. It comes in standard sizes, usually 4' x 8', 4' x 10', and 4' x 12'. It comes in varying thicknesses, from 3/8" to 3/4", in 1/8" increments. Drywall, sheetrock, and plasterboard are kinds of gypsum board.

header. A beam placed perpendicular to joists and to which joists are nailed. Used to frame an opening such as a skylight. It is also used as a lintel for a door or window.

joists. One of a series of parallel beams used to support floor and ceiling loads.

nail blocks. Pieces of wood that are installed in the framing where there is a need to have a place to nail the wall or ceiling covering to. Nail blocks are quite frequently used in the corners of the stud walls. Also called **nailers.**

nailers. See **nail block.**

nails. The two most general types are **common** nails and **finish** nails. Common nails have large flat heads, while finish nails have a small head which makes it easy to countersink. Nails range in size from 2d to 60d.

one-by-. This is a term for lumber measurement; used for any board whose width dimension is not the critical one, but whose thickness is. One-by material comes standard in most lumber yards in widths of 2" to 12", in 2" increments.

penny. The measurement for the length and corresponding diameter of a nail. The symbol for penny is a lower-case d. The larger the number, the larger the nail (e.g., 8d, 12d, 16d.)

plaster lathe. A building material made from gypsum and fastened to the frame of a building. This material usually comes in 2' x 4' sections. Plaster lathe is used as a base for a plaster wall.

plasterboard. See **gypsum board.**

plate glass. A tempered glass used for large window and door areas. It comes in a variety of thicknesses and is usually cut to size.

Plexiglas. Plexiglas is the trademark of the Rohm and Haas Co. for its acrylic plastic sheet. It is a high-impact resistant, light-weight, weather-resistant plastic. Comes in clear transparent, translucent, and opaque colors in a variety of sizes and thicknesses.

plywood. A piece of wood made of three or more layers of veneer joined with glue and usually laid with the grain of adjoining plies at right angles. The standard size of plywood is 4' x 8'; it comes in varying thicknesses, from 1/8" to 3/4", in 1/8" increments.

rafter. A structural member of the roof, designed to support roof loads.

roll insulation. Made from a blanket of mineral fiber, such as rock or glass wool or kraft paper, with side tabs for fastening to the studs. Comes in thicknesses of 2", 3", 4", and 6", and in widths of 16" or 24".

Romex. Plastic sheathed cable with solid copper conductors; can be used indoors or outdoors.

scaffold. A raised, temporary work platform made of either wood or metal.

shaft (shaftway). In skylight construction, an open space between the roof and the ceiling. Its function is to direct light to the room below. The shaftway area may be lined with any suitable wall covering.

sheathing. A covering, usually boards or plywood, used over studs or rafters.

sheetrock. See **gypsum board.**

shim. Small wedge-shaped piece of wood. It is used for filling space.

silicone caulk. A flexible sealant used to fill joints. It is unaffected by water and temperature, and has a life expectancy of 20 years.

skylight material. In this book, used to refer to the piece of translucent material (Plexiglas, plate glass, wire glass) that is set in the curb.

storm window (internal). Thin sheet of Plexiglas which is fastened to a 1" x 2" on the inside perimeter of the curb. Its purpose is to create a dead air space for insulation.

stud. One of a series of vertical structural members, used for support in walls and partitions. Studs are usually 2 x 3's or 2 x 4's.

toe-nailing. Driving a nail at a slant against the initial surface in order to permit the nail to penetrate the second surface.

top plate. The top piece of studding to which the vertical support members of the wall are nailed.

two-by- . This is a term for any piece of dimensional lumber where the width is not critical but the thickness is. Two-by material comes standard in most lumber yards in widths ranging from 3" to 12".

weather stripping. Narrow sections of a thin material, such as brass or felt, used to prevent air flow around windows and doors.

wire glass. Glass that is reinforced by large loop screening laminated between two layers of plate glass.

wood screws (W.S.). Used to fasten two parts together. These screws are threaded from the point to approximately 2/3 of the length of the screw. They have a slotted or recessed (Phillips) head for turning. Available in steel, brass, or aluminum.